# GREEN NANOARCHITECTONICS

# GREEN NANOARCHITECTONICS

edited by

**Kaushik Pal**

JENNY STANFORD
PUBLISHING

*Published by*

Jenny Stanford Publishing Pte. Ltd.
101 Thomson Road
#06-01, United Square
Singapore 307591

Email: editorial@jennystanford.com
Web: www.jennystanford.com

**British Library Cataloguing-in-Publication Data**
A catalogue record for this book is available from the British Library.

**Green Nanoarchitectonics**



ISBN 978-981-4968-50-8 (Hardcover)
ISBN 978-1-003-31860-6 (eBook)

# Contents

# Preface

This book explores the recent breakthroughs and current research trends in the design and synthesis of greener nanomaterials and their eco-friendly utilization at the industrial scale. It is multidisciplinary in nature and discusses micro- and nanofabrication, green chemistry approaches, engineered nanomaterials, spectroscopic characterization, promising possibilities for eco-friendly products, and sustainable industrial-scale applications. Comprising 12 chapters that cover diverse fields of studies, the book defines key material descriptors required for their successful employment in different applications and discusses their cost-effective synthesis from natural extracts. Such materials are desirable to synthesize integrated and adaptive materials and systems development of renewable energy technologies. The overall structure of the book is well-defined to improve its accessibility. Chapter 1 explores phytomedicines that can be improved with the help of a nanoformulation-based drug delivery system, which has great potential future for improving the therapeutic action of phytomedicines. Chapter 2 discusses the simple and cost-effective preparation of novel bionanocomposite materials, which play a significant role in nanomaterial synthesis and also act as surfactants. Chapter 3 gives an overview of the drawbacks and challenges for the photocatalytic remediation of organic pollutants such as dye molecules, emerging contaminants, and heavy metals. Chapter 4 focuses on superior therapeutic systems, including the application of green nanomaterials in targeted and sophisticated drug delivery, theranostics, regenerative medicine, implantable devices, and tissue reconstruction, and these systems are compared with conventional therapeutic system. Chapter 5 presents a systematic outline of the problem beginning with the mechanism of photocatalysis followed by the factors that affect the efficiency of a photocatalyst and various green nanomaterials used for photocatalytic activities. Chapter 6 centers around the emerging nanosorbents for water and wastewater treatment that use diverse sources of plants, microorganisms, and biocompatible green reagents for the production of green nanomaterials (GNMs), which are utilized for the decontamination and recovery of heavy metals from industrial effluents. Chapter 7 thoroughly investigates how the wood-based product industry might make use of a variety of widely available nanomaterials to improve the performance of existing products or develop new forest-based value-added products. Chapter 8 is an in-depth study of plants' hidden tendency to remodel the inorganic metal ions into nanoparticles through their gifted organic resources, which has opened up new doors to study biochemical analysis. Chapter 9 demonstrates that effective biosynthesis substantially eliminates the use of toxic reagents and employs natural resources to create nanomaterials with potential applications, implying a cost-effective and environmentally friendly solution. This chapter further highlights the improvements in this field and points out the factors that limit efficiency optimization. Chapter 10 proposes how the detailed understanding of nano/bio-interfaces would give a boost to the research on their optimal use in various disease diagnostic and therapeutic techniques. Chapter 11 depicts a comprehensive analysis focusing on the advantages, drawbacks, and prospects of organo-metallic hybrid nanomaterials synthesized especially for green and sustainable nanotechnology applications. Chapter 12 discusses the use of cosmetics with reduced impact on the environment and switching to green approach in cosmetics while not compromising with performance, retention, striking appearance, and, more important, safety. Thus, it may clear to the readers that nanotechnology-enabled green nanoarchitechtonics are starting to scale up dramatically. As they become mature and cost effective in the decades to come, nano/biomaterials could eventually replace traditional, environmentally unfriendly technologies and fossil fuels and also improve the performance of the biomedical industry through the utilization of nanocatalysts, manufacturing materials with high durability.

The book is based on the evidence from academicians, scientists, scholars, and engineers and illustrates the wide-ranging interest in the aforementioned areas. The contributions also address the novel synthesis of high-yield nanomaterials and their biomaterials, graphene, polymeric nanomaterials, green nanomaterials, green polyester, nanobiotechnology, interesting response characteristics of exclusive spectroscopic investigation as well as extensive electron microscopic study, health care, environmental and plant biology, and social, ethical, and regulatory implications of the industrial utilization of green nanotechnology–based leading functional nanomaterials. With appropriate regulation along with the topics indicated, commercial green production of manufactured nanocomposite materials can be realized.

Lastly, I would like to express my gratitude to the authors and the coauthors for their excellent research contributions.

**Kaushik Pal, PhD, DSc (Malaysia)**
Editor

## Chapter 1

# Fundamental Research Trends of Green Nanoscience and Nanotechnology

**Archana S. Patil[a] and Shailendra S. Suryawanshi[b]**
[a]*Department of Pharmaceutics, KLE College of Pharmacy, Belagavi, KLE Academy of Higher Education and Research, Belagavi 590010, Karnataka, India*
[b]*Department of Pharmaceutical chemistry, KLE College of Pharmacy, Belagavi, KLE Academy of Higher Education and Research, Belagavi 590010, Karnataka, India*
archupharma@gmail.com

Nanotechnology has received an exceptional deal of public interest because of the desire and utilization of nanomaterials in various fields of human attempts including industry, business, agriculture, public health, and medicine. Because of their unique optical, thermal, and catalytic properties, nanoparticles are employed in biomedical research, rendering it one of nanotechnology's most valuable applications. Nanomaterials are rapidly being used in biomedicine, which opens up intriguing possibilities for the creation of novel equipment and technologies across a wide range of medical fields. Despite huge commercial and public investment, concerns

*Green Nanoarchitectonics*
Edited by Kaushik Pal

ISBN 978-981-4968-50-8 (Hardcover), 978-1-003-31860-6 (eBook)
www.jennystanford.com

have been expressed about risks posed by nanomaterials. Therefore, in this chapter we present an overview of nanotechnology and its toxicity on human and ecosystem health along with different risk assessment methods. The advanced research on nanoparticle toxicity with respect to various nanomaterials have been exemplified with specific findings, responsible mechanisms and possible toxicities on human health in this chapter. Green nanotechnology is based on green chemistry (GC), or the framework of GC principles has been instrumental in the creation of innovative nanotechnologies for reciprocal economic, social, and health/environmental benefit. Thus, in addition GC and its application in green nanoscience, application of phytocompounds along with their challenges, and applications in green nanotechnology were further explored. Overall, we can conclude that the application of green nanoscience aims to minimize or eliminate risks to environment and health of society by utilizing nature's capacity to eradicate or reduce health and environmental threats associated with nanomaterials. Further, many issues with phytomedicines can be improved with the help of nanoformulation-based drug delivery system that has great potential future for improving the therapeutic action of phytomedicines.

## 1.1 An Overview of Nanotechnology and Its Toxicology

Richard P. Feynman, a Nobel Laureate in physics, first presented the notion of nanotechnology in his renowned speech "There's Plenty of Room at the Bottom" given at the American Physical Society's December 1959 conference [1]. Nanotechnology's goal is to control the forms and sizes of structures, devices, and systems at the nanoscale in order to offer intentional design, characterization, manufacturing, and application. Nanomaterials are distinguished by their size-dependent characteristics and attributes, which arise as a result of extraordinary electrical, surface activity, optical, and magnetic capabilities along with geometries of nanoscale particles [2, 3]. Electronics, water purification, food industries, information and communication technologies, agriculture, medicine, chemicals, coatings, cosmetics, and energy, all have been affected by nanotechnology [4, 5]. Companies may create new technologies that are more sustainable by integrating nanoparticles into their goods. Nanotechnology not only influences the efficiency and design of these items, but it also generates significant income [6]. Because of their small size (1–100 nm), nanoparticles are extremely helpful, as their tiny size allows them to reach a wide range of biological settings and equips them with useful size-dependent characteristics that may be used in applications. Though synthetic nanomaterials offer a wide range of uses and advantages today, their development and implementation have traditionally been costly, and in certain circumstances have resulted in the generation of ecologically harmful by-products [7].

Given the huge rates of nanomaterial synthesis, the danger of their release to the atmosphere, as well as the consequences for ecosystem health, seems to be an increasingly serious issue that must be addressed [8]. Nanostructures will get in to the environment because of both deliberate and accidental releases, including the air emissions and liquid or solid industrial waste from manufacturing plants. Nanomaterials in personal and health care, paints and textiles are also released into the environment in proportion to their usage. Nanomaterials emitted will eventually settle on earth and water surfaces. Nanomaterials that reach the earth can pollute soil and move into surface and ground waterways. Wind or rainstorm runoff can transport particles from water discharges, solid waste, direct discharges, or accidental leakages into aquatic systems.

Besides the advantages that nanostructures would enhance human lives, the technology's possible consequences are yet unknown. Do nanoparticles that have been developed represent a threat to the environment? Small particles smaller than 10 μm are respirable and can enter the alveolar areas of lungs, which is a matter of concern. Although academics seek to fill in the gaps in the information to aid risk evaluators, experts all over the world urge the need of regulating and overseeing manufacturers to reduce the hazards of nanomaterial exposure to employees, customers, and animals. To do this, it is important to comprehend the fates and functioning of produced nanosystems in the atmosphere.

Nanomaterials have the potential to affect the environment in three ways: (1) direct action on microorganisms, invertebrates, fish, and other species; (2) interactions with pollutants, which may alter toxic compound and nutrient bioavailability; and (3) changes to nonliving environmental structures [9]. Nanotoxicity is difficult to assess because of its small size and varied interactions with the environment and biological milieu. The following are some of the current nanotoxicity risk assessment approaches in use:

- Evaluation of the materials' physicochemical qualities, which allow them to interact with and potentially harm biological systems.
- Mammalian toxicity (acute and chronic tests, oral toxicity, cutaneous toxicity, skin irritation tests), mutagenicity tests, and eco-toxicity tests to allow some assessment of environmental danger.
- A set of tests for in vitro cellular assays that indicate the response of a variety of cell types after nanomaterial uptake and distribution, which can be targeted at the portal of entry or systemic sites. In general, in vitro data at the cellular level are more relevant for interpreting the mechanism of nanomaterial biokinetics. The results obtained in vitro can be gathered to predict in vivo Absorption, Distribution, Metabolism and Excretion (ADME)/toxicity of the nanomaterials through systematic information on (a) the effective cellular uptake and bioavailability at target sites, (b) cellular metabolism and organ toxicity, and (c) cellular excretion and tissue accumulation and long-term risks.
- Exposure assessment involves both qualitative and quantitative information on the duration, concentration, frequency, and material of exposure of humans or the environment.
- Hazard identification, to understand what are the adverse health and environmental effects associated with particular nanoparticles [10].

The solutions to these and many other issues will drive the development of regulatory standards that will safeguard the environment while also allowing nanotechnology's benefits to be fully realized.

## 1.2 Various Nanomaterials and Toxicities

Due to advancements in nanotechnology, industrial nanomaterials have numerous remarkable chemical and physical characteristics; their usage in many sectors is being studied throughout the world. Nanosystems produce a huge range of environmental toxicity in various ways. Few nanomaterials, such as carbon nanotubes (CNTs), quantum dots (QDs), metal nanoparticles, magnetic nanomaterials, and others, have substantial toxicities. CNTs are utilized in a variety of sectors, including optical instruments, solar cell mobiles, semiconductors, capacitors, and space elevator cables with unique properties [11, 12]. A quantum dot is a type of nanoparticle that has been widely studied. QDs exhibit unique optical, semiconductor, photochemical, and catalytic characteristics due to quantum confinement phenomena. Conventional QDs are promising for biological applications due to surface changes to heavy metal QDs and the use of heavy metal-free QDs [13]. Silica nanoparticles possess wide range of characteristics and thus exhibit a huge variety of applications. They are abrasive, tough compounds that effectively reduce friction and are thus used to paint waxed floors and even train tracks. Because of their absorptive characteristics, they can be used as a drainage aid in papermaking. Rubber, polymers, and concrete can all benefit from their use as a binding agent. Most importantly, they are nontoxic and stable materials with a huge range of applications in biomedical, medication delivery, and optical imaging agents [14]. Also, anticancer, medication delivery, radiotherapy enhancement, thermal ablation, diagnostic tests, antibacterial, antifungal, gene delivery, and many other biomedical applications have proven that metal- and metal oxide–supported nanomaterials have a substantial therapeutic impact [15, 16]. Apart from the above-mentioned advantages of nanotechnology, there are indeed several risks connected with this innovative technology. Because nanoparticles are easily absorbed by biological systems, there are worries regarding nanotechnology's impact on humans. Table 1.1 summarizes the reported mechanisms and potential risks associated with nanoparticles.

**Table 1.1** Various nanomaterials, mechanisms, and their possible risk factors

| Nanomaterials | Mechanisms responsible | Possible risk factors | References |
|---|---|---|---|
| Carbon nanotubes | When respirable particles are breathed into the lungs and phagocytized by alveolar macrophages, inflammatory cytokines as well as chemokines are formed, and recurring exposure to respirable particles in the lungs causes chronic inflammatory damage. Due to excess or improper healing mechanisms, chronic inflammatory damage eventually leads to fibrosis of lungs and respiratory cancer. | Pulmonary inflammation and injury, tumors in the respiratory system are common, genotoxicity. | 17–20 |
| Quantum dots | The toxicity of QDs mainly depends on individual QDs' diameter, charge, content, exterior functional groups; photolytic, oxidative, and mechanical stability of QDs, as well as their physicochemical characteristics and environmental circumstances. | Toxic effects due to skin penetration | 21 |

| Nanomaterials | Mechanisms responsible | Possible risk factors | References |
|---|---|---|---|
| Silica nanoparticles | The cytotoxicity of SiNPs is determined by the particle's physicochemical characteristics. They alter macrophage/monocyte activity and ability, increase antigen-specific cellular immune responses, and alter cell functioning. Dysfunction of autophagy. | Induced immunotoxicity | 22 |
| Iron oxide magnetic nanoparticles | The chemical composition, size, and dose of the substance, as well as its accumulation in the body, specific toxicity to organs, immunogenicity, metabolism, and elimination from the body. | Inflammation, and reductions in growth rate, ulceration, survival, and neurobehavioral changes in plants and cell lines, as well as animal models. | 23–25 |
| Silver and gold nanoparticles | Redistribution and accumulation in vital organs, the emission of silver ions that cause biochemical changes. Toxicities also depend on special characters of nanomaterials, namely, size, surface engineering, and dose. | Neurotoxicity, genotoxicity, inflammation, oxidative stress and interference with intracellular signaling | 26–29 |

(*Continued*)

**Table 1.1** (*Continued*)

| Nanomaterials | Mechanisms responsible | Possible risk factors | References |
|---|---|---|---|
| $MnO_2$ nanomaterials | Dissolution of $MnO_2$ inside the lysosomal compartment, suppression of respiration (basal and maximum), and the supplementary respiratory ability of gill cells are all possible processes along with different biochemical and biodistribution profiles of diverse tissues. | Dysfunctioning of mitochondria and declined cellular respiratory function, genotoxicity | 30, 31 |
| $TiO_2$ nanoparticles | Production of reactive oxygen species (ROS), and skin penetration | Oxidative stress, genotoxicity, metabolic variations, inflammation, as well as carcinogenesis. | 11, 32 |
| $Al_2O_3$, carbon black, Co, and Ni nanoparticles | Physicochemical characteristics of particles, namely, nanosize and surface area are responsible. | High pulmonary toxicity compared to micron-sized particles | 33, 34 |

Nanotechnology's use in several sectors is restricted because of its toxicological consequences on human health and the environment. To avoid such negative effects, scientists are currently attempting to manufacture nanomaterials using green methods. Green nanoscience, as GC, aims to minimize or eliminate risks to environment and health of society by utilizing nature's capacity to eradicate or reduce health and environmental threats associated with nanomaterials.

## 1.3 GC and Its Principles

Green chemistry is the branch of chemistry that mainly deals with the "the utilization of a set of principles that minimizes or eliminates the use or generation of hazardous substances in the design, manufacture, and application of chemical products." We have total 12 principles in GC, given by Warnerand Anastas [35]. The principles of GC nowadays are mainly applied to develop, design, and manufacture different products including medicines and electronic devices. The objectives of these principles are to reduce the waste, prevent pollutions, and decrease the hazards caused by reagents and chemicals. It also helps to use less amount of chemicals, reagents, and solvents during production of different kinds of materials [36]. Application of these principles has reduced the use of hazardous reagents and solvents. They also help in the improvement of material efficiency and its quality for safer use [37].

The principles of 12 GC are very important and essential for fundamental and sustainable development of society with the advancement in scientific areas. The GC principles are described in Table 1.2.

- The main areas that adopt the principles of GC include [38]:
- New chemical synthesis, processes, and methodologiesDrug and product development

**Table 1.2** Green chemistry principles

| S. no. | Principles | Description |
|---|---|---|
| 1 | "Better to prevent than to cure" | It is recommended to avoid the formation of waste material before starting any work instead of cleaning up and treating waste later. |
| 2 | "Atom economy" | To achieve atom economy, suitable synthetic pathways should be planned to use maximum amount of raw materials and precursor compounds to form desired product. |

*(Continued)*

**Table 1.2** (*Continued*)

| S. no. | Principles | Description |
|---|---|---|
| 3 | "Less precarious chemical syntheses" | During the synthesis or production of any product or chemical entity, we need to plan and follow suitable methods and synthetic routes that may yield maximum product quantity and decrease the quantity hazardous waste. |
| 4 | "Designing safer chemicals" | Products or chemical agents with required quality, functionality should be prepared by considering the toxicity of product or chemicals. |
| 5 | "Safer solvents and safer auxiliaries" | The cost of auxiliary agents like reagents, separating liquids, solvents, chemicals should be avoided in maximum possible cases. It is recommended to utilize auxiliaries and agents that are harmless. |
| 6 | "Design for energy efficiency" | The efficiency of energy is very much important during production of material that affects economic and environmental factors. The chemical synthesis and methodologies should be analyzed and optimized using required energy inputs. The chemical processes are recommended to perform using ambient temperature, pressure, and mild conditions. |
| 7 | "Renewable feedstocks" | Whenever feasible in technological and economic terms, the synthetic methods should adopt the use of raw materials and chemicals and feedstocks, which are not limited and renewable. |
| 8 | "Derivative reduction" | In the synthesis and production of chemicals, it is recommended not to utilize additional steps and chemicals that are responsible to generate extra chemical waste. The steps like derivatization of functional groups, protection, de-protection and blocking agents should be avoided or minimally used. |

| S. no. | Principles | Description |
|---|---|---|
| 9 | "Catalysis" | The uses of catalytic reagents are recommended in the chemical reactions. The use of stoichiometric and selective reagents is acceptable. |
| 10 | "Degradation" | The chemical entities should be planned and designed in such a way that after the life span and use, the synthesized chemicals should not resist in the biosphere and they should be easily degraded into nonhazardous waste. |
| 11 | "Real-time analysis for pollution prevention" | To prevent the pollution, there is need of developing advanced instrumentation techniques for the real-time analysis, monitoring of in-line methods, and control of such steps before the formation of hazardous waste. |
| 12 | "Accident prevention by inherently safer chemistry" | To reduce the risk and accidents, fire formation in the laboratory, it is recommended to use the suitable compounds and chemical methodologies. |

### 1.3.1 Utility of Principles of GC

- The principles of GC are approached to develop nanomaterials, nanomedicines, and nanodevices that can be added into high-performance products that are less hazardous to society and environment [39].
- It helps in discovering the synthetic procedures that use the reagents that are less harmful and also increase the efficiency of existing methodologies [37].
- The utilization of GC principles toward nanoscience help to facilitate the formation of safer nanodevices and materials.
- Many applications of green nanoscience and nanotechnology make use of principles of GC in order to prepare the nanoscale materials [40].
- The principles of GC are helpful to give proactive design which assures that the production of nanomaterials is done in safer way and also it is done by considering the biological and ecological hazards [41].

- It plays a very important role in the field of nanoscience that helps in the maximum benefits to the society and reduces the impact on ecosystem [37].
- GC guides for the production of materials, development of new method, and design of application throughout the life cycle [42, 43].

### 1.3.2 Green Nanoscience to Apply Principles of GC

The green nanoscience is a very important branch of science that deals with applications of green principles of chemistry and green engineering. It mainly helps to reduce the utility of fuel and energy by employing the minimum use of materials and renewable inputs wherever needed. The well-being of society and health of population is the final and very important goal of any type of development in scientific areas. To achieve this goal, implementation of new strategies is needed. Well-being of population can be achieved by providing effective therapeutic medicines and treatments; hence, safe delivery of medicines is important [44]. The clean and green technology development is a rapidly growing field nowadays. The green nanoscience, in phytochemical-based preparations or phytomedicines, significantly play a very essential role in maintaining environmental sustainability with respect to nanoformulation by avoiding any type of harm to environment, society, and health of the human [45].

The utilization of plant materials and related phytomedicines in the preparation of nanoparticles and nanoformulation is essential, rational, and recommended due to its easy accessibility and availability. The usefulness of phytomedicines is accepted in the nanopreparations due to the large variety of primary and secondary metabolites. Diverse number of biological activity shown by the phytochemicals include antioxidant, antibacterial, antiviral, anticancer, antidiabetic activities, and hence, they are greatly involved in the nanopreparations for effective therapeutic management of diseases and disorders. In the green nanoscience and green nanotechnology, the word "green" refers to the utilization of phytomaterials or phytomedicines. The green nanoscience mainly uses the principles of GC and it helps in the energy minimization and fuel utilization. Many nanotechnological preparations, processes, and utilizations are found to possess the high energy utilization, large amount of fuel utilization, large utilization of water, reduces the greenhouse gases, and generates the hazardous waste. To overcome the limitations of nanotechnology, the concept of green nanoscience and nanotechnology is evolved. The green nanoscience is the approach to minimize the adverse reactions or effects before production [46, 47].

## 1.4 A Green Approach in the Development of Nanoparticles

Phytomedicines are large source of drugs and the biological activities of such medicines are due to the combined effect of active components of that plant material. The phytoconstituents present in the herbal material show synergistic activity, which in turn improves their therapeutic effect. Phytomedicines are very important and play essential role in the management of various serious diseases and disorders. But the major limitations or disadvantages of phytomedicines include:

- Solubility issues
- Low bioavailability
- Difficulty in absorption
- Difficulty in systemic clearance
- Frequent administration of drug
- Higher dose utilization
- Difficulty in reaching the site of action.

To overcome above limitation of phytomedicines or preparations based on herbal origin, a new research area has been evolved in the phytoformulation research known as nanoformulation. This new area mainly deals with development and validation of nanotechnology-based dosage formulations of drugs from natural origin. Few of the very important nanotechnology-based herbal formulations include polymeric nanoparticles, nanocapsules, nanospheres, proliposomes, nanoemulsions, solid lipid nanoparticles, and liposomes. These novel formulations are very important in the field of phytopharmacological research and advantages of same are presented in Fig. 1.1.

Hence, many issues with phytomedicines can be improved with the help of "nanoformulation-based drug delivery system" (NDDS). This has great potential future for improving their therapeutic action of phytomedicines. The help of nanocarriers as an "NDDS" in traditional drug delivery systems is needed to manage chronic diseases like cancer, diabetes, asthma, and others with the utilization of phytomedicines [48–50].

**Figure 1.1** Advantages of nanoformulations prepared from herbal medicines.

Many important phytomedicines-loaded nanoparticles have been developed and effectively used in the management of chronic diseases such as cancer, diabetes, and hypertension. Various drugs obtained from plants, which have been loaded into nanoformulation, include curcuminoids, glycyrrhizin, flavonoids, lignans, artemisinin, berberin, camptothecin, and taxel. Table 1.3 summarizes the phytomedicines-loaded nanoformulations [51–57].

**Table 1.3** Phytomedicines-loaded nanoformulations along with therapeutic effect and method of preparation

| S. no. | Active phytoconstituents | Type of nanoformulation | Therapeutic effect | Preparation method |
|---|---|---|---|---|
| 1 | Artemisinin | Nanocapsules | Anticancer activity | Ionic gelation method |
| 2 | Berberine | Berberine-loaded nanoparticles | Anticancer activity | Self-assembly procedure |
| 3 | Camptothecin | Camptothecin encapsulated nanoparticles | Anticancer activity | Dialysis method |
| 4 | Curcuminoids | Curcuminoids solid lipid nanoparticles | Anticancer and antioxidant activity | Microencapsulation method |
| 5 | Glycyrrhizin | Glycyrrhizin-loaded nanoparticles | Anti-inflammatory and anti-hypertensive activity | Rotary-evaporated film ultrasonication method |
| 6 | Flavonoids and lignans | Nanoparticles of *Cuscuta chinensis* | Anti-oxidant and hepatoprotective activity | Nanosuspension method |
| 7 | Taxel | Taxel-loaded nanoparticles | Anticancer activity | Emulsion solvent evaporation method |

## 1.5 Applications or Trends of Green Nanoscience and Nanotechnology

The concept of green nanoscience and nanotechnology is mainly applicable and implemented in huge areas like, bioengineering, nanofabrication, cosmetics, energy, drugs and medicines, nano-biotechnology, and optical engineering [26]. Various research trends of nanotechnology and nanoscience are explained as below and represented in Fig. 1.2.

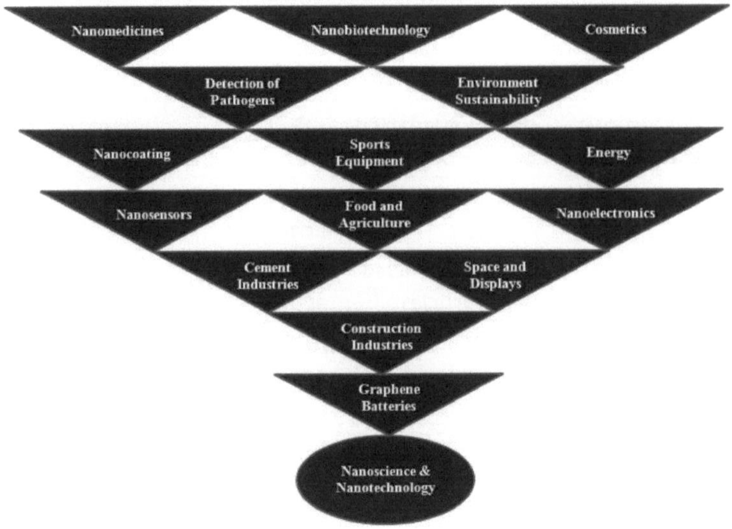


**Figure 1.2** Applications and nanoscience and nanotechnology.

### 1.5.1 Nanomedicine

The utilization of nano-based drug delivery is most potential in the management of various metabolic and other chronic diseases and different types of disorders. The principles of GC and its applications in technology and science for treating, diagnosing, preventing disease and disorders, relieving pain, injury, and improving health of society using molecular tools and knowledge of the human body can be done by using the nanoscience and nanotechnology. The comprehensive control, construction, defense, management, monitoring, and improvement of all human biological systems are useful tools of nanoscience [58].

### 1.5.2 Nanobiotechnology

Nanobiotechnology is the important technique which utilizes the principles of nanoscience and biotechnology. Nanobiotechnology has vital and great role in the development and application of various tools in the life science studies. Implementation of nanomaterials with that of the biological principles leads to development of different types of analytical tools, treatments, therapy, diagnostic devices, contrast agents, drug delivery vehicles [59].

### 1.5.3 Nanotechnology in Cosmetics

The applications of nanotechnology, nanoscience, and nanomaterials are very much helpful in the production, cosmetic preparation such as moisturizers, hair care products, sunscreens, and make-up preparations [60].

### 1.5.4 Detection of Foodborne Illnesses

The principles of "nanoscience and nanotechnology" are helpful in the identification techniques for different types of pathogens. Techniques and materials, such as dye immobilized nanoparticles, fluorescence, luminescence, and metallic nanoparticles, are the useful tools in detecting pathogenic diseases. Nanotechnology helps to detect rapid, sensitive, reliable, and simple isolation and identification method for pathogens [61].

### 1.5.5 Nanotechnology and the Environment

The climate and environment protection can be very effectively done by utilizing the principles of green nanoscience and nanotechnology. During the production of various nanotechnological equipments, apparatus, devices, products, and methods, principles of GC can be applied, which are definitely helpful in reducing the use of hazardous substances, minimizing the use of raw chemicals, reagents and materials, also energy minimization, and water utilization.

### 1.5.6 Nanotechnology in Sports Equipment

It has many benefits and strength to improve sports equipments that make athletes comfortable, safer, and more agile. For example, the durability and performance of various sports equipments, such as tennis, baseball bats, hockey sticks, badminton racquets, racing bicycles, golf balls, skis, archery, and arrows, can be improved.

### 1.5.7 Food and Agriculture

The techniques and principles of nanoscience and nanotechnology are useful in the production of food items and agricultural materials and devices. It has potential impact and applications in the preparation of functional food, and manufacturing of other biological food products.

### 1.5.8 Nanocoatings

Nanocoating is very important in the field of aerospace, marine, medical, and oil industries. It is important in the production of multifunctional coatings to different products. The nanocoating is a type of coating that involves the application of nanoscale thin films to surfaces of materials, which helps to improve the function of material, and prevent from corrosion.

### 1.5.9 Energy

The nanotechnology innovations are very important in different types of energy sectors, energy sources, energy storage, energy distribution, energy conversion, and energy utilization. Nanoscience and nanotechnologies help to increase the efficiency of energy in almost all types of industries.

### 1.5.10 Nanosensors

The nanosensors are sensing materials with dimensions smaller than 100 nm and they are mainly used for collecting data from different sources. They are developed for the identification of gases from chemical, physical, and biochemical variables, and also for the detection radiations of electromagnetic waves.

### 1.5.11 Nanoelectronics

It is a very important branch of electronics dealing with the use and implications of nanotechnology in electronic devices. These electronic components are having very small size, may be ranging from few nanometers. Nanoelectronics mainly cover the materials and devices with very small size with physical effects that alter the properties of materials on nanoscale quantum mechanical and inter atomic interactions.

### 1.5.12 Nanotechnology in Furniture

The principles of nanoscience and technology are helpful in the production of furniture, which helps to decrease the need for functional textiles and adhesives. Nanoscience is helpful in producing good quality of furniture with different properties and applications.

### 1.5.13 Graphene Batteries

The nanoscience and technology are very much helpful in the production of graphene batteries. The graphene is chemically processed and used for manufacturing of different types of graphene batteries with ultrahigh energy. It is helpful in the production of various types of batteries such as redox flow, lithium, sulfur, metal, and air.

### 1.5.14 Nanotechnology in Space

Nanoscience and nanotechnology has important role in future space missions. "Nanosensors, dramatically improved high-performance materials, or highly efficient propulsion systems can be produced by using principles of nanoscience."

### 1.5.15 Nanotechnology in the Automotive Industry

The automotive industry is a largest material-based industry and nanotechnology help to improve the performance of existing technologies. The nanoscience and nanotechnology has useful impact in the generation of lighter but stronger materials, fuel cells, production of good paint quality, batteries, ultra-thin anti-glare layers for windows and mirrors to the futuristic energy harvesting bodywork, switchable colors, fully self-repairing paint, and shape shifting skin.

### 1.5.16 Nanotechnology in the Construction Industry

Nanotechnology has a great role and impact in the construction industries. It helps in improving the performance and durability of various components used in construction. This also helps in the safety and energy efficiency of the buildings.

### 1.5.17 Nanotechnology in the Cement Industry

In cement industry, it has a potential role to address issues such as poor crack resistance, carbon dioxide emissions along with low tensile strength, curing time, absorption of high water, and many other mechanical properties. The nanoengineering of cement-based preparations are much helpful and they are outstanding in their performance and also has smart properties.

## 1.6 Conclusions

Normally, during the period of advancement of new nanotechnology area, experts concentrate essentially around recognizing new properties and applications. Subsequently, the assessment of any accidental properties of the material (e.g., environmental or health hazards) or concerns about risks or efficiencies of the production cycle is regularly conceded. Assessment of the potential toxicological and ecological impacts of nanoscale materials before they are acknowledged as full-grown advancements presents a chance to limit putative unfortunate results from the beginning and eventually lead to the plan of better materials. The phytoconstituents present in the natural material give synergistic action which helps in the improvement of restorative impact of medication, however, they have restricted clinical applications. Blend of green science and nanoscience can be possibly applied to defeat the restrictions of phytoformulations and also proved to be beneficial in developing production-level commercial scale materials. Further many issues with phytomedicines can be improved with the help of nanoformulation-based drug delivery systems that has great potential future for improving the therapeutic action of phytomedicines. Commercialization will rely heavily on the development of high-precision, low-waste nanomanufacturing techniques. GC, in addition to improving research and development methods, has the potential to improve public opinion of nanoscience since, it is very simple to explain and may be utilized to express a proactive approach towards the development of this new technology. For these factors, GC can play a significant role in directing the development of nanotechnology to ensure that these products benefit society and the environment to the greatest extent possible. Further development and application of green nanoscience to the design and production of various nanomaterials will provide research opportunities and challenges for the foreseeable future.

## References

1. Feynman R.P. There's plenty of room at the bottom [data storage]. *Journal of Microelectromechanical Systems*, 1992, **1**(1): 60–66.
2. Bhainsa K.C., D'souza S.F. Extracellular biosynthesis of silver nanoparticles using the fungus *Aspergillus fumigatus*. Colloids and Surfaces B: Biointerfaces, 2006, **47**(2): 160–164.
3. Shahverdi A.R., Fakhimi A., Shahverdi H.R., Minaian S. Synthesis and effect of silver nanoparticles on the antibacterial activity of different antibiotics against *Staphylococcus aureus* and *Escherichia coli*. Nanomedicine: Nanotechnology, Biology and Medicine, 2007, **3**(2): 168–171.
4. Yang F., Deng D., Pan X., Fu Q., Bao X. Understanding nano effects in catalysis. *National Science Review*, 2015, **2**(2): 183–201.
5. Robinson I. Nanotechnology and green technology: how can the two work hand in hand? AZoNano, https://www.azonano.com/article.aspx?ArticleID=5017.
6. Drexled K.E. *Engines of Creation: The Coming Era of Nanotechnology*, 1986 (Oxford: Oxford University Press).
7. Lim B., Jiang M., Camargo P.H., Cho E.C., Tao J., Lu X., Zhu Y., Xia Y. Pd-Pt bimetallic nanodendrites with high activity for oxygen reduction. *Science*, 2009, **324**(5932): 1302–1305.
8. Ray C.P., Yu H., Fu PP. Toxicity and environmental risks of nanomaterials. *Journal of Environmental Science and Health, Part C: Environmental Carcinogenesis & Ecotoxicology Reviews*, 2009, 27(1): 1–35.
9. Viswanath B., Kim S. Influence of nanotoxicity on human health and environment: the alternative strategies. *Reviews of Environmental Contamination and Toxicology*, 2017, 242: 61–104.
10. Khan S.H., Fulekar M.H., Pathak B. Nanotoxicology-health and environmental impacts: a review. *Journal of Environmental Nanotechnolgy*, 2015, **4**(3): 55–72.
11. Bonn P.J., Driscoll K. Particles, inflammation and respiratory tract carcinogenesis. *Toxicology Letters*, 1996, **88**(1–3): 109–13.
12. Shacter E., Weitzman S.A. Chronic inflammation and cancer. *Oncology*, 2002, **16**(2): 217–226.

13. Cotta M.A. Quantum dots and their applications: what lies ahead? *ACS Applied Nano Materials*, 2020, **3**(6): 4920–4924.

14. Jeelani P.G., Mulay P., Venkat R., Ramalingam C. Multifaceted application of silica nanoparticles: a review. *Silicon*, 2020, **12**(6): 1337–1354.

15. Venkatesh N., Bhowmik H., Kuila A. Metallic nanoparticle: a review. *Biomedical Journal of Scientific & Technical Research*, 2018, **4**(2): 3765–3775.

16. Yaqoob A.A., Ahmad H., Parveen T., Ahmad A., Oves M., Ismail I.M., Qari H.A., Umar K., Mohamad Ibrahim M.N. Recent advances in metal decorated nanomaterials and their various biological applications: a review. *Frontiers in Chemistry*, 2020, 8: 341.

17. Nishi K., Morimoto Y., Ogami A., Murakami M., Myojo T., Oyabu T., Kadoya C., Yamamoto M., Todoroki M., Hirohashi M., Yamasaki S. Expression of cytokine-induced neutrophil chemoattractant in rat lungs by intratracheal instillation of nickel oxide nanoparticles. *Inhalation Toxicology*, 2009, **21**(12): 1030–1039.

18. Ogami A., Morimoto Y., Myojo T., Oyabu T., Murakami M., Nishi K., Kadoya C., Tanaka I. Histopathological changes in rat lung following intratracheal instillation of silicon carbide whiskers and potassium octatitanate whiskers. *Inhalation Toxicology*, 2007, **19**(9): 753–758.

19. Bonn P.J., Driscoll K. Particles, inflammation and respiratory tract carcinogenesis. *Toxicology Letters*, 1996, **88**(1–3): 109–113.

20. Shacter E., Weitzman S.A. Chronic inflammation and cancer. *Oncology*, 2002, **16**(2): 217–226.

21. Hardman R. A toxicologic review of quantum dots: toxicity depends on physicochemical and environmental factors. *Environmental Health Perspectives*, 2006, **114**(2): 165–172.

22. Chen L., Liu J., Zhang Y., Zhang G., Kang Y., Chen A., Feng X., Shao L. The toxicity of silica nanoparticles to the immune system. *Nanomedicine*, 2018, **13**(15): 1939–1962.

23. Ran Q., Xiang Y., Liu Y., Xiang L., Li F., Deng X., Xiao Y., Chen L., Chen L., Li Z. Eryptosis indices as a novel predictive parameter for biocompatibility of $Fe_3O_4$ magnetic nanoparticles on erythrocytes. *Scientific Reports*, 2015, **5**(1): 1–5.

24. Malhotra N., Chen J.R., Sarasamma S., Audira G., Siregar P., Liang S.T., Lai Y.H., Lin G.M., Ger T.R., Hsiao C.D. Ecotoxicity assessment of $Fe_3O_4$ magnetic nanoparticle exposure in adult zebrafish at an environmental pertinent concentration by behavioral and biochemical testing. *Nanomaterials*, 2019, **9**(6): 873.

25. Malhotra N., Lee J.S., Liman R.A., Ruallo J.M., Villaflores O.B., Ger T.R., Hsiao C.D. Potential toxicity of iron oxide magnetic nanoparticles: a review. *Molecules*, 2020, **25**(14): 3159.

26. Báez D.F., Gallardo-Toledo E., Oyarzún M.P., Araya E., Kogan M.J. The influence of size and chemical composition of silver and gold nanoparticles on in vivo toxicity with potential applications to central nervous system diseases. *International Journal of Nanomedicine*, 2021, 16: 2187.

27. Li Y.F., Chen C. Fate and toxicity of metallic and metal-containing nanoparticles for biomedical applications. *Small*, 2011, **7**(21): 2965–2980.

28. Katsnelson B.A., Privalova L.I., Gurvich V.B., Makeyev O.H., Shur V.Y., Beikin Y.B., Sutunkova M.P., Kireyeva E.P., Minigalieva I.A., Loginova N.V., Vasilyeva M.S. Comparative in vivo assessment of some adverse bioeffects of equidimensional gold and silver nanoparticles and the attenuation of nanosilver's effects with a complex of innocuous bioprotectors. *International Journal of Molecular Sciences*, 2013, **14**(2): 2449–2483.

29. Bondarenko O., Juganson K., Ivask A., Kasemets K., Mortimer M., Kahru A. Toxicity of Ag, CuO and ZnO nanoparticles to selected environmentally relevant test organisms and mammalian cells in vitro: a critical review. *Archives of Toxicology*, 2013, **87**(7): 1181–1200.

30. Browning C.L., Green A., Gray E.P., Hurt R., Kane A.B. Manganese dioxide nanosheets induce mitochondrial toxicity in fish gill epithelial cells. *Nanotoxicology*, 2021, **15**(3): 400–417.

31. Singh S.P., Kumari M., Kumari S.I., Rahman M.F., Mahboob M., Grover P. Toxicity assessment of manganese oxide micro and nanoparticles in Wistar rats after 28 days of repeated oral exposure. *Journal of Applied Toxicology*, 2013, **33**(10): 1165–1179.

32. Grande F., Tucci P. Titanium dioxide nanoparticles: a risk for human health? *Mini Reviews in Medicinal Chemistry*, 2016, **16**(9): 762–769.

33. Oberdörster G., Ferin J., Lehnert B.E. Correlation between particle size, in vivo particle persistence, and lung injury. *Environmental Health Perspectives*, 1994, **102**(Suppl. 5): 173–179.

34. Warheit D.B., Webb T.R., Colvin V.L., Reed K.L., Sayes C.M. Pulmonary bioassay studies with nanoscale and fine-quartz particles in rats: toxicity is not dependent upon particle size but on surface characteristics. *Toxicological Sciences*, 2007, **95**(1): 270–280.

35. Kargozar S., Ramakrishna S., Mozafari M. Chemistry of biomaterials: future prospects. *Current Opinion in Biomedical Engineering*, 2019, 10: 181–190.

36. Valavanidis A., Vlachogianni T., Fiotakis K. Laboratory experiments of organic synthesis and decomposition of hazardous environmental chemicals following green chemistry principles. In *International Conference "Green Chemistry and Sustainable Development"*, Thessaloniki, 2009, pp. 25–26.

37. Anastas P.T., Warner J.C. *Green Chemistry: Theory and Practice*, 1998 (New York: Oxford University Press).

38. Dahl J.A., Maddux B.L., Hutchison J.E. Toward greener nanosynthesis. *Chemical Reviews*, 2007, **107**(6): 2228–2269.

39. Kharissova O.V., Kharisov B.I., Oliva González C.M., Méndez Y.P., López I. Greener synthesis of chemical compounds and materials. *Royal Society Open Science*, 2019, **6**(11): 191378.

40. Iavicoli I., Leso V., Ricciardi W., Hodson L.L., Hoover M.D. Opportunities and challenges of nanotechnology in the green economy. *Environmental Health*, 2014, **13**(1): 1.

41. Geraci C., Heidel D., Sayes C., Hodson L., Schulte P., Eastlake A., Brenner S. Perspectives on the design of safer nanomaterials and manufacturing processes. *Journal of Nanoparticle Research*, 2015, **17**(9): 1–3.

42. Mohammed T.I. Green chemistry as ecofriendly chemistry: a review. *International Journal*, 2020, **1**(01): 1–4.

43. Hullmann A., Meyer M. Publications and patents in nanotechnology. *Scientometrics*, 2003, **58**(3): 507–527.

44. Zou H., Wu S., Shen J. Polymer/silica nanocomposites: preparation, characterization, properties, and applications. *Chemical Reviews*, 2008, **108**(9): 3893–3957.

45. Balbus J.M., Florini K., Denison R.A., Walsh S.A. Protecting workers and the environment: an environmental NGO's perspective on nanotechnology. *Journal of Nanoparticle Research*, 2007, **9**(1): 11–22.

46. Verma A., Gautam S.P., Bansal K.K., Prabhakar N., Rosenholm J.M. Green nanotechnology: advancement in phytoformulation research. *Medicines*, 2019, **6**(1): 39.

47. Besley J.C., Kramer V.L., Priest S.H. Expert opinion on nanotechnology: risks, benefits, and regulation. *Journal of Nanoparticle Research*, 2008, **10**(4): 549–558.

48. Guo L., Liu X.Y., Sanchez V., Vaslet C., Kane A.B., Hurt R.H. A window of opportunity: designing carbon nanomaterials for environmental safety and health. *Materials Science Forum*, 2007, **544**: 511–516

49. Hristozov D., Ertel J., TechnoValuation M. Nanotechnology and sustainability: benefits and risks of nanotechnology for environmental sustainability. *Forum der Forschung*, 2009, **22**: 161–168.

50. Jayapalan A.R., Lee B.Y., Kurtis K.E. Can nanotechnology be 'green'? Comparing efficacy of nano and microparticles in cementitious materials. *Cement and Concrete Composites*, 2013, 36: 16–24.

51. Som C., Wick P., Krug H., Nowack B. Environmental and health effects of nanomaterials in nanotextiles and facade coatings. *Environment International*, 2011, **37**(6): 1131–1142.

52. Rickerby D.G., Morrison M. Nanotechnology and the environment: a European perspective. *Science and Technology of Advanced Materials*, 2007, **8**(1–2): 19.

53. Hutchison J.E. Greener nanoscience: a proactive approach to advancing applications and reducing implications of nanotechnology. *ACS Nano*, 2008, **2**(3): 395–402.

54. Lam C.W., James J.T., McCluskey R., Arepalli S., Hunter R.L. A review of carbon nanotube toxicity and assessment of potential occupational and environmental health risks. *Critical Reviews in Toxicology*, 2006, **36**(3), 189–217.

55. Schulte P.A., McKernan L.T., Heidel D.S., Okun A.H., Dotson G.S., Lentz T.J., Geraci C.L., Heckel P.E., Branche C.M. Occupational safety and health, green chemistry, and sustainability: a review of areas of convergence. *Environmental Health*, 2013, **12**(1): 1–9.

56. Eastlake A., Hodson L., Geraci C., Crawford C. A critical evaluation of material safety data sheets (MSDSs) for engineered nanomaterials. *Journal of Chemical Health & Safety*, 2012, **19**(5), 1–8.

57. Boulaiz H., Alvarez P.J., Ramirez A., Marchal J.A., Prados J., Rodríguez-Serrano F., Perán M., Melguizo C., Aranega A. Nanomedicine: application areas and development prospects. *International Journal of Molecular Sciences*, 2011, **12**(5): 3303–3321.

58. Surendiran A., Sandhiya S., Pradhan S.C., Adithan C. Novel applications of nanotechnology in medicine. *Indian Journal of Medical Research*, 2009, **130**(6): 689–701.

59. Katz L.M., Dewan K., Bronaugh R.L. Nanotechnology in cosmetics. *Food and Chemical Toxicology*, 2015, 85: 127–137.

60. Kumar H., Kuča K., Bhatia S.K., Saini K, Kaushal A., Verma R., Bhalla T.C., Kumar D. Applications of nanotechnology in sensor-based detection of foodborne pathogens, Sensors, 2020, **20**(7): 1966.

61. Rickerby D.G., Morrison M. Nanotechnology and the environment: a European perspective. *Science and Technology of Advanced Materials*, 2007, **8**(1–2): 19.

# Chapter 2

# Sources of Green Nanomaterials

**A. Manjceevan and K. Velauthamurty**
*Department of Chemistry, Faculty of Science, University of Jaffna, Jaffna, Sri Lanka*
manjceevan@univ.jfn.ac.lk

Synthesis of nanomaterials mediated by biological materials—termed "green nanomaterials synthesis"—has received attention due to its eco-friendly synthetic process. Nanomaterials are defined as materials with at least one of their dimensions around 1–100 nm and they could be synthesized with zero-, one-, two-, and three-dimensional materials such as quantum dots, nanotubes, nanoplates, and nanoflowers. Bionanomaterials synthesis is mediated by biological materials such as plant extract, algae, fungi, and yeast. These biomaterials play a significant role in nanomaterial synthesis and mainly act as a surfactant and reducing agent. The use of biomaterials in nanomaterial synthesis is a cost-effective process and reduces the usage of harmful, toxic chemicals. Through the green nanomaterials synthesis method, different sizes, morphologies of nanomaterials with desired properties could be synthesized and it is applicable in several applications, including photovoltaic, catalytic,

*Green Nanoarchitectonics*
Edited by Kaushik Pal

ISBN 978-981-4968-50-8 (Hardcover), 978-1-003-31860-6 (eBook)
www.jennystanford.com

biomedical, and biosensing applications. This chapter mainly summarizes the green nanomaterials synthesis and the mechanism involved in the process together with recently reported examples.

## 2.1 Overview of Nanomaterials

Research on nanomaterial synthesis received great interest due to its attractive properties that differ with morphologies, size, functional groups on nanomaterial surface; crystalline nature; and its vast number of applications. Materials with at least one of their dimensions, particularly below 100 nm on nanometer scale are referred to as nanomaterials. Nanomaterials could be classified based on their morphology such as nanoparticles, nanorod, nanotube, nanosheet, and nanoflower [1]. More interestingly, the properties of nanomaterials vary with their dimension to a great extent. Synthesized nanomaterials could be applicable in many fields including medicine, chemistry, physics, biology, agriculture, and industries. Particularly, it can be used in biosensors, catalysts, photocatalysts, fabrication of solar cells, drug synthesis, nonfertilizer, nanofillers, etc. [2–5]. However, the applications of nanomaterials have several limitations such as related to stability, defects, bioaccumulation, recycle and regeneration, use of expensive chemicals and reagents, lack of understanding in synthesis, and toxicity of chemicals. These drawbacks open the paths to perform new researches. Scientists surge to synthesize novel materials with attractive properties for desired applications toward industrial and commercial level together with safer, cleaner, environmentally friendly, and long durable products [6].

## 2.2 Properties of Nanomaterials

Nanomaterials could be classified based on the overall shape as zero-, one-, two-, and three-dimensional nanomaterials. The size of the materials plays a significant role in the physicochemical properties of nanomaterials including optical properties and electrochemical properties [7]. Nanomaterials show a high surface area per volume ratio and tend to form aggregates or agglomerates. The aggregate implies that the material exists together with strong interaction.

The material existence together with weak interaction is called an agglomerate. The surfactant that exists around the nanomaterials prevents the aggregation and agglomeration of nanoparticles [8, 9].

The high surface area over volume ratio of nanomaterial leads to many physicochemical properties and opens the path toward plenty of applications such as catalyst, medicine, and photovoltaic applications. Nanomaterial in an aqueous medium creates surface charge due to protonation and deprotonation of surface atoms. The charges on the nanomaterial surface create the electrical double layer. Surface charges are balanced by the counter ions available in the solution and it creates an electrical double layer. Surface charge is commonly measured in terms of zeta potentials. Zeta potentials indicate the electrostatic repulsion between adjacent molecules in the dispersion. If the zeta potential exists ±30 mV for nanomaterial is subject to stability in the suspension and the surface charge prevent the agglomeration of nanomaterial. Further, zeta potential between +10 mV and –10 mV is considered neutral and zeta potential more than +30 or less than –30 is referred to as strong cationic and strong anionic, respectively [10].

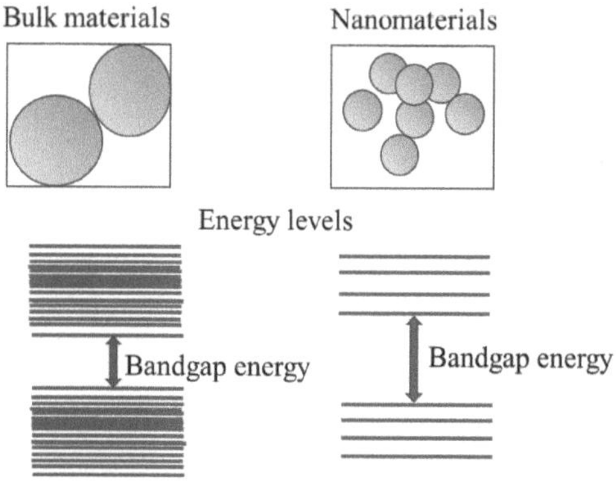


**Figure 2.1** Bandgap of nanomaterials and bulk.

Nanomaterials show different physicochemical properties than the materials in bulk and it is termed as quantum confinement effect. The quantum confinement effect is described in terms of bandgap energy, valance band position, conduction band position, and size of nanomaterials. The reduction of the particle size to the nanoscale discrete the energy levels and increase the bandgap energy as shown in Fig. 2.1. It leads to a change in the optical and electrochemical properties of nanomaterials [9]. Several studies related to nanoparticle synthesis with different sizes and morphologies by using a green synthesis approach with different plant extracts have been reported [11–13].

## 2.3 Synthesis of Nanomaterials

Nanomaterials with diverse morphology could be synthesized by numerous methods. Synthesized nanomaterials could be categorized mainly in two approaches, namely, top-down and bottom-up approaches (Fig. 2.2).

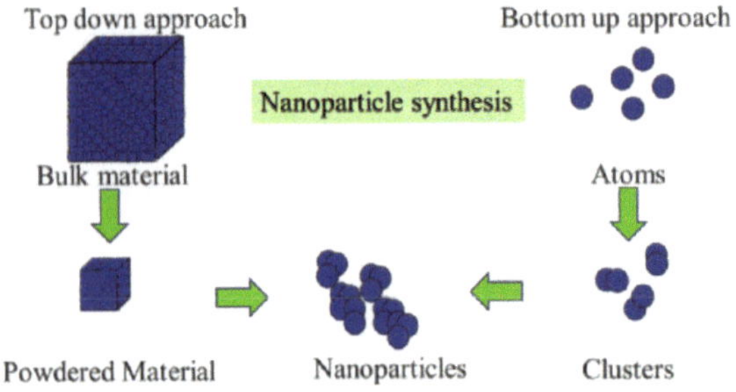


**Figure 2.2** Different approaches of nanoparticles synthesis.

In the top-down approache, starting materials break down as small particles and it includes high-energy ball milling, sputtering, laser ablation, electro-explosion, and lithography (Fig. 2.3).

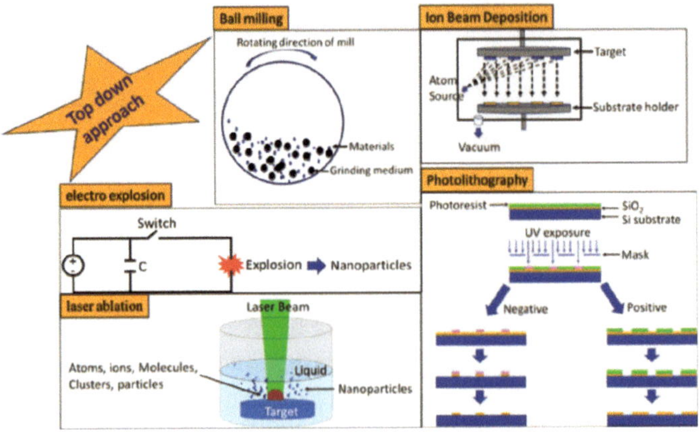


**Figure 2.3** Top-down approaches of nanomaterial synthesis.

Synthesis of nanomaterials from atomic or molecular species is referred to as the bottom-up approach. It includes ultrasonic synthesis, sol-gel methods, colloidal synthesis, hydrothermal approaches, solvothermal approaches, and green synthesis (Fig. 2.4).

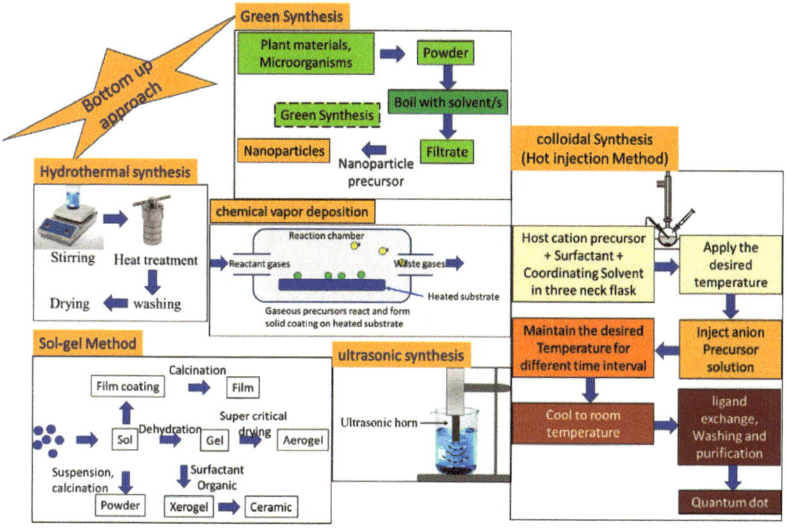


**Figure 2.4** Bottom-up approaches of nanomaterial synthesis.

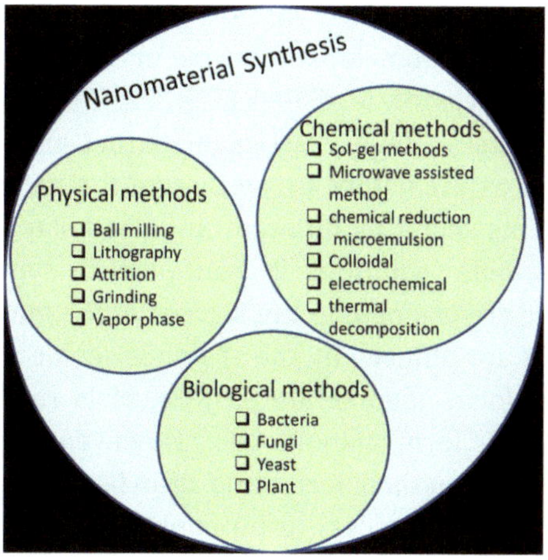


**Figure 2.5** Classification of nanomaterial synthesis.

The method of nanoparticle synthesis could be classified as physical, chemical, and biological synthesis (Fig. 2.5). The physical synthesis includes lithography, ball milling, and vapor-based synthesis [14]. Chemical nanoparticle synthesis includes sol-gel methods, chemical reduction methods, microwave-assisted methods, colloidal synthesis, and electrochemical methods. In the chemical method, the variety of toxic chemicals used as reducing agents and solvents limits the applications. Commonly, citric acid, polyethylene glycol, polyvinylpyrrolidone, polyvinyl alcohol, ethylenediaminetetraacetic acid, oleylamine, trioctylphosphine, and chitosan are used as capping agents [15–17]. Sodiumborohydrate, sodium citrate, and l-ascorbic acid are used as reducing agents [18, 19]. The nanomaterials synthesis by using biological materials is called the green synthesis of nanomaterials and it eliminates many obstacles associated with chemical methods. The green synthesis of nanomaterials is a nontoxic, clean, eco-friendly method.

## 2.4 Green Chemistry Synthetic Approach of Nanomaterials

The green synthesis method is an environmentally friendly process in which one or more harmful chemicals are substituted by the extract of biological material. More clearly, the green synthesis could be explained that the synthesis process reduces/prevention of waste/toxic emission, substitute one or more components with environmentally friendly feedstock (Fig. 2.6) [6]. For green synthesis approaches, bacteria, fungi, yeast, plant extract, and waste materials could be used as mediators to synthesize the nanomaterials. The biological agents secret by different microbes play differently with the different metal solutions in nanoparticle synthesis. Initially, the metal ions are trapped on the surface of the microbes and then the metal ions are reduced by the enzymes secret by the particular microbes and form nanometals [20]. Bacteria remarkably reduce metal ions and form nanometals. However, the nanoparticle synthesis by this method has some drawbacks such as bacterial laboratory culturing process, little control over morphology, size shape, and distribution. Fungi produce intracellular and extracellular enzymes and produce nanoparticles with well-defined sizes and morphologies. Several fungal species are used in nanoparticle synthesis such as *Penicillium chrysogenum, Fusarium oxysporum, Fusarium chlamydosporum, Verticillium luteoalbum, Trichothecium* sp., *Aspergillus oryzae, Aspergillus tamarii, Colletotrichum* sp., *Alternaria alternata*, and *Trichoderma viride* [20, 21]. Nanoparticle production using fungi also has some limitations such as being time-consuming and the need to find a cost-effective method for commercial-scale production [22]. However, the plant-mediated nanosynthesis has several benefits over the use of microbes such as safe handle, rapid rate of synthesis, most stable nanoparticle, better control over size, shape, and morphology, suitable for large-scale synthesis. [14]. Plant extract consists mainly of phytochemicals that play a major role in nanomaterial production. Mainly flavanols and bioactive phenols play a major role in nanomaterial synthesis. Plant extract consisting of carbohydrates and proteins act as reducing agents and form metal nanoparticles. However, the mechanism behind the metal oxides nanoparticle synthesis is still unknown. Researchers suggest that the metal ions reduced by phytochemical are present in the plant extract and then binding with the oxygen either exists in the atmosphere or in degrading phytochemical and form oxides. The formed oxides are stabilized by phytochemical and form nano-oxides [22].

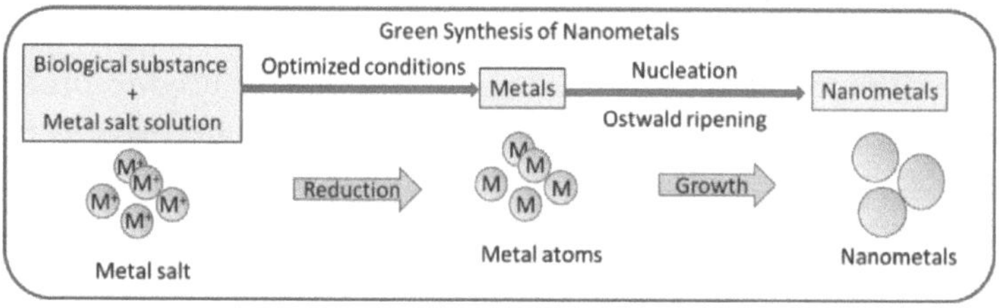


**Figure 2.6** Schematic illustration of green nanomaterial synthesis.

The advantages of green synthesis of nanomaterials are environmentally friendly processes, energy-saving, and cost-effective methods. The use of biological materials as surfactants offers several benefits such as prevention of the aggregation/agglomeration of nanomaterials, yield of biocompatible nanomaterials, and application to several biomedical fields as catalysts, nanobiosensor, optical imaging and labeling of the biological system, antimicrobial activity, and biomedical diagnostics [6]. By the green synthesis method, various shapes and sizes of nanomaterials could be produced with better characteristic features. The properties of such green synthesis nanomaterials could be tuned by the synthesis of nanomaterials in various conditions such as temperature, pressure, pH, solvent, surfactant, and concentration of precursor [23]. Plant materials consist of phytochemicals, polyphenols, carotenoids, ketones, aldehyde, flavonoids, and terpenes [23, 24]. Plant materials may act as surfactants and reducing agents to reduce metal ions to produce metal nanoparticles. Chemical and green synthesis methods of nanomaterial synthesis required a solvent system. Water is an ideal, cheap, and commonly accessible solvent for the synthesis of materials. In green nanomaterials synthesis, water or natural plant extract is used as solvent system. Green synthesis of nanomaterials has plenty of benefits including cost-effective, eco-friendly methods, possible to perform in one pot synthesis and need low energy to initiate the reaction [6]. Microbes play a significant role in nanomaterial synthesis due to their nature reduces toxic ions to detoxified through reduction by the enzymes and proteins that exist in intracellular and extracellular microbes [25].

### 2.4.1 Bacteria-Mediated Nanomaterials Synthesis

Nonpathogenicity of probiotic microorganisms is used to produce nanomaterial and it is an eco-friendly, cost-effective process, involving living cells, and produces stable nanomaterials than a conventional chemical method of nanomaterial synthesis [26]. Those grow rapidly and produce proteins and enzymes and are excellent candidate used for nanoparticle synthesis. Probiotics attract cations due to their negative electro kinetics potential [26]. It reduces the metal ions as metal [27]. Bacteria and acetamides are widely used for metal and metal oxide synthesis [6]. Bacteria is used for the synthesis of various nanoparticles of different sizes and morphologies. The nanomaterials production mediated by probiotic bacteria is tabulated in Table 2.1.

**Table 2.1** Probiotic bacteria used to produce nanomaterials

| Nanomaterial | Probiotic bacteria | Properties of nanomaterials | References |
|---|---|---|---|
| Au NPs | *Deinococcus radiodurans* | Spherical, triangular, and irregular shapes with an average size of 43.75 nm | 28 |
| Au NPs | *Acinetobacter* sp. | Spherical, monodispersed crystalline AuNPs | 29 |
| ZnO NPs | *Lactobacillus sporogens* | The average particle size of 145.7 nm | 30 |
| ZnO NPs | *Aeromonas hydrophila* | Spherical, oval shape, an average size of 57.72 nm | 31 |
| CdS | *Lactobacillus* sp. *and Sachharomyces cerevisiae* | Individual nanoparticles, few aggregates, size of 2.5–5.5 nm | 32 |
| $PbS_2$ | *Idiomarina* sp. | Tetragonal crystal lattice, a crystallite domain size of 2.38 nm, and an interplanar distance of 0.318 nm | 33 |
| Cu NPs | *Shewanella loihica* | Size in the range of 10–16 nm | 34 |
| $Fe_3O_4$ NPs | *Bacillus subtilis* | 70 nm | 35 |
| Se NPs | *Bacillus cereus* | 170 nm | 36 |
| Ag, AgCl NPs | *Raoultella planticola* and *Pantoea agglomerans* | Mean particle size of around 10–50 nm | 37 |
| CdS | *E. coli* | 2–5 nm | 38 |

### 2.4.2 Fungi-Mediated Nanomaterial Synthesis

Fungi is a nonphototropic microorganism that has a rigid cell wall. Fungi consist of a large number of enzymes, glycoproteins, and polysaccharides on the surface including glucan and chitin [14]. In the same condition, fungi grow quickly than bacteria and the fungi consist of a higher surface area as they consist of large mycelia [25]. Enzymes and proteins in fungi act as reducing agents, reduce metal ions, and produce nanoparticles to a great extent than bacteria. It consists of various intracellular enzymes used to synthesize metal and metal oxide nanomaterials. Reducing agents that exist in fungi reduce metal salt directly and produce metal atoms. Metal ions are converted as metal nanoparticles intra- and extracellularly. The extracellular route performs much rapidly than the intracellular route of nanomaterial synthesis. The surface of fungi trapping the metal ions and the enzymes and protein secret by fungi reduces it as metals [25]. The reduction of each metal ion reduces by specific fungal enzymes and proteins, like nitrate reductase essential to the reduction of ferric ions [25]. There are several pieces of research performed to synthesize the nanoparticle mediated by fungi (Table 2.2).

**Table 2.2** Nanomaterial production mediated by fungi

| Nanomaterial | Fungi | Mode of action fungi | Properties of nanomaterials | References |
|---|---|---|---|---|
| $TiO_2$ NPs | *Aspergillus flavus* | Reducing and capping agent | Spherical, oval in shape. Size of 62–74 nm | 39 |
| Ag NPs | *Verticillium* | Intracellular reduction | 25 ± 12 nm | 40 |
| Au NPs | *Alternaria alternata* | Extra cellular | Average particle size of 12 ± 5 nm | 41 |
| ZnO NPs | *Aspergillus fumigatus* | Extracellular secretions | Nanoparticle size ranging between 1.2 and 6.8 nm at least in one dimension with oblate spherical and hexagonal in structure | 42 |

| Nanomaterial | Fungi | Mode of action fungi | Properties of nanomaterials | References |
|---|---|---|---|---|
| Au NPs | *Penicillium brevicompactum* | Extracellular | 20 to 50 nm, spherical gold nanoparticles, triangular- and hexagonal shaped particles | 43 |
| Ag NPs | *Anamorphous Bjerkandera* | Extracellular | 10 and 100 nm | 44 |
| Ag NPs | *Raphanus sativus* | – | Particles size of 4–30 nm, spherical | 45 |
| Cu and CuO NPs | *Stereum hirsutum* | Extracellular protein | Mainly spherical, 5 to 20 nm | 46 |

### 2.4.3 Yeast-Mediated Nanomaterial Synthesis

Yeast is a eukaryotic single-cell microorganism. It is classified under the kingdom fungi. More than 1500 yeast species were identified [6]. Nanoparticles and nanomaterials were synthesized successfully by using yeast extract and this is published in several articles (Table 2.3).

**Table 2.3** Nanomaterial production mediated by yeast

| Nanomaterial | Fungi | Properties of nanomaterials | References |
|---|---|---|---|
| CdS NPs | *Candida glabrata* and *Schizosaccharomyces pombe* | Intracellular particles of diameter 20 Å | 47 |
| Ag and Au NPs | *Extremophilic yeasts* | Ag NPs smaller than 20 nm, AuNPs diameter ranged from 30 to 100 nm | 48 |
| Ag NPs | *Dry baker's yeast* | Uniform spherical shape and fine size, with an average size of 13.8 nm | 49 |

(*Continued*)

**Table 2.3** (*Continued*)

| Nanomaterial | Fungi | Properties of nanomaterials | References |
|---|---|---|---|
| $TiO_2$ NPs | *Baker's yeast* | Spherical in shape with an average size of 6.7 ± 2.2 nm | 50 |
| ZnO NPs | *Pichia kudriavzevii* | Hexagonal wurtzite structure and a particle crystal size of ~10 to 61 nm | 51 |

## 2.4.4 Plant Extract Used for Nanomaterials Synthesis

Different morphologies of nanostructures were successfully synthesized by plant extract–mediated nanosynthesis. Biomolecules such as proteins, amino acids, polysaccharides, enzymes, and organic acid involved in the plant extract mediated bionanosynthesis [14]. These biomolecules act as capping and reducing agents. Plant-mediated biosynthesis has many potential applications such as cost-effective method, safe to handle, eco-friendly, better control over size and shape, suitable for large-scale nanosynthesis, one-pot synthesis, and rapid process [14]. Plant materials could be used in powder form or as an extract. The metal ions are mixed with plant material or extract and allowed to form nanomaterials in optimized process conditions such as optimized temperature, pH, and concentration of precursor. The functional groups that exist in the biomolecules such as C=C, C=O, O–H, COOH groups reduce the metal ions as metals [14]. Plant materials such as stem, root, leaf, latex, flower, seeds, fruit are being used for the biosynthesis of nanomaterials. Several nanometals and nanometal oxides are synthesized by plant-mediated syntheses such as silver, gold, platinum, palladium, copper, and other metals and metal oxides including copper oxide, titanium oxide, zinc oxide synthesis, etc. The synthesis of nanomaterials by plant extract–mediated process are tabulated in Table 2.4.

**Table 2.4** Nanomaterial synthesis mediated by different plant extracts

| Nano-materials | Plant extract | Medium of extract | Functional groups involved | Properties of nano-materials | References |
|---|---|---|---|---|---|
| Ag NPs | *Tropaeolum majus* L | Methanolic/ aqueous extracts | –C=O, –O–H | Cylindrical shape with size of 25 μm | 52 |
| Ag NPs | *Acalypha wilkesiana* | Aqueous | –OH, –C=C, –C–H, $–CH_3$, –C–N | Spherical size between 10 and 26 nm | 53 |
| Au NPs | *Salix alba* L. | Aqueous | Hydroxyl group and glucosidal linkages | Particle size of 63 nm (50–80 nm) | 54 |
| Pt NPs | *Iraqi Zahidi* dates | Aqueous | Polyphenols | Spherical shapes (30–45) nm | 55 |
| MgO | *Prosopis farcta* | Aqueous | –OH groups, C=C, | Nanowire | 56 |
| Pt NPs | *Prosopis farcta (P. farcta) fruits* | Aqueous | Aliphatic ($sp^3$), carbonyl and hydroxyl groups and aromatic ($sp^2$) | Face-centered cube (fcc) crystal, mean diameter of 3.8 nm | 57 |
| Pd NPs | *Fritillaria imperialis* flower | Aqueous | C–O, C=C and O–H | 10–20 nm | 58 |
| S doped $TiO_2$ | *Malva parviflora* | Aqueous | Smaller average particle size (20.3 vs. 29.5 nm) with a larger surface area (135 vs. 95 $m^2/g$) | 20.3 nm, larger surface area (135 $m^2/g$) | 59 |

## 2.5 Conclusion and Future Trends

The biosynthesis of nanoparticles gained significant attention due to its benefits such as biocompatibility, low-cost approach, eco-friendly process, and one-pot simple process and so on. The vast variety of biological materials widely used in these syntheses such as plant extract, and microorganisms includes bacteria, fungi, yeast, etc. Silver, gold, platinum, palladium, $TiO_2$, ZnO, and several composites were synthesized by using green synthesis methods. In this process, the biological components existing in the extract such as proteins, enzymes, polysaccharides, flavanone, terpenes, and others consist of the functional groups involved as capping agents and reducing agents to synthesize the nanomaterials. The capping agents stabilize the nanoparticles and prevent the aggregation and agglomeration of nanomaterials. However, the synthesis procedure depends on several factors, including temperature, pH and precursor concentration, and content of active ingredients in the extract as well. The mechanism behind the biosynthesis proposed is based on the analysis of functional groups. However, different plant extracts have different functional groups, and the existence of active components also varied widely. Therefore, more attention should be paid to elucidate the mechanisms precisely of green nanosynthesis using different plant extracts by using sophisticated techniques.

## References

1. Khan, I., K. Saeed, and I. Khan, Nanoparticles: Properties, applications and toxicities. *Arabian Journal of Chemistry*, 2019, **12**(7), 908–931.
2. Manjceevan, A., Quantum-dot-based photoelectrochemical biosensors: principles, fabrication, and applications, in Bio-Manufactured Nanomaterials: Perspectives and Promotion, K. Pal, editor, *Springer International Publishing, Cham*, 2021, pp. 23–40.
3. Barhoum, A., et al., Nanofibers as new-generation materials: From spinning and nano-spinning fabrication techniques to emerging applications. *Applied Materials* Today, 2019, **17**, 1–35.
4. Manjceevan, A., N. Sulaimalebbe, and T. Somapala, Visible-light-harvesting hedgehog like copper bismuth oxide: Optical, structural and electrochemical properties. *Journal of Molecular Structure*, 2021, **1235**, 130205.

5. Manjceevan, A. and J. Bandara, Robust surface passivation of trap sites in PbS q-dots by controlling the thickness of CdS layers in PbS/CdS quantum dot solar cells. *Solar Energy Materials and Solar Cells*, 2016, **147**, 157–163.
6. Singh, J., et al., 'Green' synthesis of metals and their oxide nanoparticles: applications for environmental remediation. *Journal of Nanobiotechnology*, 2018, **16**(1), 84.
7. Gatoo, M.A., et al., Physicochemical properties of nanomaterials: Implication in associated toxic manifestations. *BioMed Research International*, 2014, **2014**, 498420.
8. Biener, J., et al., Surface chemistry in nanoscale materials, *Materials* (Basel), 2009, **2**(4), 2404–2428.
9. Asha, A.B. and R. Narain, Chapter 15: Nanomaterials properties, in *Polymer Science and Nanotechnology*, R. Narain, editor, Elsevier, *Amsterdam*, 2020, pp. 343–359.
10. Clogston, J.D. and A.K. Patri, Zeta potential measurement, in *Characterization of Nanoparticles Intended for Drug Delivery*, S.E. McNeil, editor, Humana Press, Totowa, NJ, 2011, pp. 63-70.
11. Akhtar, M.S., J. Panwar, and Y.-S. Yun, Biogenic synthesis of metallic nanoparticles by plant extracts. *ACS Sustainable Chemistry & Engineering* 2013, **1**(6), 591–602.
12. Si, A., et al., Sustainable preparation of gold nanoparticles via green chemistry approach for biogenic applications. *Materials Today Chemistry*, 2020, **17**, 100327.
13. Makarov, V.V., et al., Biosynthesis of stable iron oxide nanoparticles in aqueous extracts of hordeum vulgare and rumex acetosa plants. *Langmuir*, 2014, **30**(20), 5982–5988.
14. Vijayaraghavan, K. and T. Ashokkumar, Plant-mediated biosynthesis of metallic nanoparticles: A review of literature, factors affecting synthesis, characterization techniques and applications. *Journal of Environmental Chemical Engineering*, 2017, **5**(5), 4866–4883.
15. Javed, R., et al., Role of capping agents in the application of nanoparticles in biomedicine and environmental remediation: Recent trends and future prospects. *Journal of Nanobiotechnology*, 2020, **18**(1), 172.
16. Dheyab, M.A., et al., Simple rapid stabilization method through citric acid modification for magnetite nanoparticles. *Scientific Reports*, 2020, **10**(1), 10793–10793.
17. Villaverde-Cantizano, G., et al., *Reducing Agents in Colloidal Nanoparticle Synthesis–an Introduction*. Royal Society of Chemistry, 2021: pp. 1–27.

18. Hussain, M.H., et al., Synthesis of various size gold nanoparticles by chemical reduction method with different solvent polarity. *Nanoscale Research Letters*, 2020, **15**(1), 140.
19. Yokoyama, S., et al., Green synthesis of Cu micro/nanoparticles for low-resistivity Cu thin films using ascorbic acid in aqueous solution. *Journal of Materials Chemistry C*, 2016, **4**(31), 7494–7500.
20. Fariq, A., T. Khan, and A. Yasmin, Microbial synthesis of nanoparticles and their potential applications in biomedicine. *Journal of Applied Biomedicine*, 2017, **15**(4), 241–248.
21. Ghosh, S., et al., Mechanistic aspects of microbe-mediated nanoparticle synthesis. *Frontiers in Microbiology*, 2021. **12**(867): 638068.
22. Jeevanandam, J., Y.S. Chan, and M.K. Danquah, Biosynthesis of metal and metal oxide nanoparticles. *ChemBioEng*, 2016, **3**(2), 55–67.
23. Sharma, D., S. Kanchi, and K. Bisetty, Biogenic synthesis of nanoparticles: A review. *Arabian Journal of Chemistry*, 2019, **12**(8), 3576–3600.
24. Nasrollahzadeh, M., et al., Chapter 3: Biological sources used in green nanotechnology, in *Interface Science and Technology*, M. Nasrollahzadeh, et al., editors, Elsevier, Amsterdam, 2019, pp. 81–111.
25. Siddiqi, K.S. and A. Husen, Fabrication of metal nanoparticles from fungi and metal salts: Scope and application. *Nanoscale Research Letters*, 2016, **11**(1), 98.
26. Bahrulolum, H., et al., Green synthesis of metal nanoparticles using microorganisms and their application in the agrifood sector. *Journal of Nanobiotechnology*, 2021, **19**(1), 86.
27. Iravani, S., Bacteria in nanoparticle synthesis: Current status and future prospects. *International Scholarly Research Notices*, 2014, **2014**, 359316.
28. Li, J., et al., Biosynthesis of gold nanoparticles by the extreme bacterium Deinococcus radiodurans and an evaluation of their antibacterial properties. *International Journal of Nanomedicine*, 2016, **11**, 5931.
29. Nadhe, S.B., et al., Green synthesis of AuNPs by Acinetobacter sp. GWRVA25: Optimization, characterization, and its antioxidant activity. *Frontiers in Chemistry*, 2020, **8**, 474.
30. Mishra, M., et al., Studies on the inhibitory activity of biologically synthesized and characterized zinc oxide nanoparticles using lactobacillus sporogens against Staphylococcus aureus. *Journal of Pure and Applied Microbiology*, 2013, **7**(2), 1263–1268.

31. Jayaseelan, C., et al., Novel microbial route to synthesize ZnO nanoparticles using Aeromonas hydrophila and their activity against pathogenic bacteria and fungi. Spectrochimica Acta Part A: *Molecular and Biomolecular Spectroscopy*, 2012, **90**, 78–84.

32. Prasad, K. and A.K. Jha, Biosynthesis of CdS nanoparticles: An improved green and rapid procedure. *Journal of Colloid and Interface Science*, 2010, **342**(1), 68–72.

33. Srivastava, P. and M.J.A. Kowshik, Fluorescent lead (IV) sulfide nanoparticles synthesized by Idiomarina sp. strain PR58-8 for bioimaging applications. *Applied and Environmental Microbiology*, 2017, **83**(7), e03091–16.

34. Lv, Q., et al., Biosynthesis of copper nanoparticles using Shewanella loihica PV-4 with antibacterial activity: Novel approach and mechanisms investigation. *Journal of Hazardous Materials*, 2018, **347**, 141–149.

35. Sundaram, P.A., et al., Extracellular biosynthesis of iron oxide nanoparticles by Bacillus subtilis strains isolated from rhizosphere soil. *Biotechnology and Bioprocess Engineering* 2012, **17**(4), 835–840.

36. Pouri, S., et al., Biological synthesis of selenium nanoparticles and evaluation of their bioavailability. *Journal of Brazilian Archives of Biology Technology*, 2017. **60**.

37. Ghiuta, I., et al., Bacteria-mediated synthesis of silver and silver chloride nanoparticles and their antimicrobial activity. *Journal of Applied Sciences*, 2021, **11**(7), 3134.

38. Sweeney, R.Y., et al., Bacterial biosynthesis of cadmium sulfide nanocrystals. *Chemical Biology*, 2004, **11**(11), 1553–1559.

39. Rajakumar, G., et al., Fungus-mediated biosynthesis and characterization of TiO2 nanoparticles and their activity against pathogenic bacteria. *Spectrochimica Acta Part A: Molecular and Biomolecular Spectroscopy*, 2012, **91**, 23–29.

40. Mukherjee, P., et al., Fungus-mediated synthesis of silver nanoparticles and their immobilization in the mycelial matrix: A novel biological approach to nanoparticle synthesis. *Nano Letters*, 2001, **1**(10), 515–519.

41. Sarkar, J., et al., Mycogenesis of gold nanoparticles using a phytopathogen Alternaria alternata. *Bioprocess and Biosystems Engineering*, 2012, **35**(4), 637–643.

42. Raliya, R. and J.C. Tarafdar, ZnO nanoparticle biosynthesis and its effect on phosphorous-mobilizing enzyme secretion and gum contents in

clusterbean (Cyamopsis tetragonoloba L.). *Agricultural Research*, 2013, **2**(1), 48–57.

43. Mishra, A., et al., Microbial synthesis of gold nanoparticles using the fungus Penicillium brevicompactum and their cytotoxic effects against mouse mayo blast cancer C 2 C 12 cells. *Applied Microbiology and Biotechnology*, 2011, **92**(3), 617–630.
44. Osorio-Echavarría, J., et al., Synthesis of silver nanoparticles using white-rot fungus Anamorphous Bjerkandera sp. R1: Influence of silver nitrate concentration and fungus growth time. *Scientific Reports*, 2021, **11**(1), 3842.
45. Singh, T., et al., Biosynthesis, characterization and antibacterial activity of silver nanoparticles using an endophytic fungal supernatant of Raphanus sativus. *Journal of Genetic Engineering and Biotechnology*, 2017, **15**(1), 31–39.
46. Cuevas, R., et al., Extracellular biosynthesis of copper and copper oxide nanoparticles by Stereum hirsutum, a native white-rot fungus from Chilean forests. *Journal of Nanomaterials*, 2015, **2015**, 789089.
47. Dameron, C.T., et al., Biosynthesis of cadmium sulphide quantum semiconductor crystallites. *Nature*, 1989, **338**(6216), 596–597.
48. Mourato, A., et al., Biosynthesis of crystalline silver and gold nanoparticles by extremophilic yeasts. *Bioinorganic Chemistry and Applications*, 2011, **2011**, 546074.
49. Shu, M., et al., Biosynthesis and antibacterial activity of silver nanoparticles using yeast extract as reducing and capping agents. *Nanoscale Research Letters*, 2020, **15**(1), 14.
50. Peiris, M., et al., $TiO_2$ nanoparticles from Baker's yeast: A potent antimicrobial. *Journal of Microbiology and Biotechnology*, 2018, **28**(10), 1664–1670.
51. Moghaddam, A.B., et al., Biosynthesis of ZnO nanoparticles by a new Pichia kudriavzevii yeast strain and evaluation of their antimicrobial and antioxidant activities. *Molecules (Basel, Switzerland)*, 2017, **22**(6), 872.
52. Bawazeer, S., et al., Green synthesis of silver nanoparticles using Tropaeolum majus: Phytochemical screening and antibacterial studies. *Green Processing and Synthesis*, 2021, **10**(1), 85–94.
53. Dada, A.O., et al., Silver nanoparticle synthesis by Acalypha wilkesiana extract: phytochemical screening, characterization, influence of operational parameters, and preliminary antibacterial testing. *Heliyon*, 2019, **5**(10), e02517.

54. Islam, N.U., et al., Green synthesis and biological activities of gold nanoparticles functionalized with Salix alba, 2019. *Arabian Journal of Chemistry*, **12**(8), 2914–2925.

55. Ali, N.H. and A.M. Mohammed, Biosynthesis and characterization of platinum nanoparticles using Iraqi Zahidi dates and evaluation of their biological applications. *Biotechnology Reports*, 2021, **30**, e00635.

56. Abdullah, O.H. and A.M. Mohammed, Biosynthesis and characterization of MgO nanowires using Prosopis farcta and evaluation of their applications. *Inorganic Chemistry Communications*, 2021, **125**, 108435.

57. Jameel, M.S., et al., Green sonochemical synthesis platinum nanoparticles as a novel contrast agent for computed tomography. *Materials Today Communications*, 2021, **27**, 102480.

58. Veisi, H., et al., Bio-inspired synthesis of palladium nanoparticles fabricated magnetic Fe3O4 nanocomposite over Fritillaria imperialis flower extract as an efficient recyclable catalyst for the reduction of nitroarenes. *Scientific Reports*, 2021, **11**(1), 4515.

59. Helmy, E.T., et al., Novel green synthesis of S-doped TiO2 nanoparticles using Malva parviflora plant extract and their photocatalytic, antimicrobial and antioxidant activities under sunlight illumination. *Chemosphere*, 2021, **271**, 129524.

# Chapter 3

# Hybrid Three-Dimensional (3D) Graphene Architectures for Photocatalysis of Noxious Pollutants

**Alvin Lim Teik Zheng**[a] **and Yoshito Andou**[b, c]
[a]*Department of Science and Technology, Faculty of Humanities, Management and Science, Universiti Putra Malaysia, Bintulu Sarawak Campus, 97008 Bintulu, Sarawak, Malaysia*
[b]*Graduate School of Life Science and Systems Engineering, Kyushu Institute of Technology, Kitakyushu, Japan*
[c]*Collaborative Research Centre for Green Materials on Environmental Technology, Kyushu Institute of Technology, Kitakyushu, Japan*
yando@life.kyutech.ac.jp

The importance of pollutant remediation using solar energy is increasingly acknowledged as a viable method to solve the current environmental problems. To date, three-dimensional (3D) graphene (Gr) architectures have demonstrated excellent photocatalytic performance offering efficient separation of photogenerated charge carriers. The facile hydrothermal treatment has been found to be the commonly reported synthetic protocol in the preparation of various novel hybrid 3D Gr architectures. This chapter focuses on the recent progress in the development and application of hybrid 3D Gr for the photocatalytic remediation of organic pollutants such as

*Green Nanoarchitectonics*
Edited by Kaushik Pal

ISBN 978-981-4968-50-8 (Hardcover), 978-1-003-31860-6 (eBook)
www.jennystanford.com

dye molecules, emerging contaminants, and heavy metals. Various novel design based on semiconductor/3D Gr has led to the creation of highly efficient heterojunction photocatalyst. In addition, we have also made an effort to discuss some photocatalytic anti-bacterial disinfection innovations. Finally, prospects in the construction of hybrid 3D Gr possessing enhanced photocatalytic materials are presented. This write-up should provide a comprehensive overview of hybrid 3D Gr architecture in wastewater remediation applications.

### Abbreviations

| | |
|---|---|
| AA | Ascorbic acid |
| BPA | Bisphenol A |
| CNTs | Carbon nanotubes |
| DCP | 2,4 dichlorophenol |
| Gr | Graphene |
| GA | Graphene aerogel |
| GO | Graphene oxide |
| NP | Nanoparticle |
| rGO | Reduced graphene oxide |
| rGH | Reduced graphene hydrogel |
| ROS | Reactive oxygen species |
| TC | Tetracycline |
| UV | Ultraviolet |
| vdW | Van der Waals |
| Vis | Visible light |
| 3D | Three-dimensional |
| 2D | Two-dimensional |

## 3.1 Introduction

Photocatalysis is among the leading green strategy in wastewater remediation contaminated with pollutants such as synthetic dye, heavy metals, emerging contaminants, and pathogens. The working principle of photocatalytic reaction is often based on the generation of holes and electrons due to irradiation from visible/ultraviolet (UV) light sources; depending on the bandgap of the photocatalysts. Graphene (Gr) and its derivatives have become ubiquitous materials in various applications extending to environmental remediation [1–5].

Owing to their salient attributes such as possessing high surface area, good electron mobility, high stability, and biocompatibility, pristine Gr has transformed the way how researcher capitalizes it for usage in many novel innovations. However, the preparation of large-scale pristine Gr requires complicated and complex synthetic protocol requiring sophisticated protocol, incurring a high cost. Previous studies have mainly focused on two-dimensional (2D) photocatalytic material based on reduced Gr oxide (rGO) or Gr oxide (GO), in which photoactive materials were anchored on their surface. The drawbacks include significant agglomeration of the sheets after the drying process due to the van der Waals (vdW) forces, inherently reducing the surface area of the hybrid photocatalysts and significantly restricted mass and photoexcited electron transport.

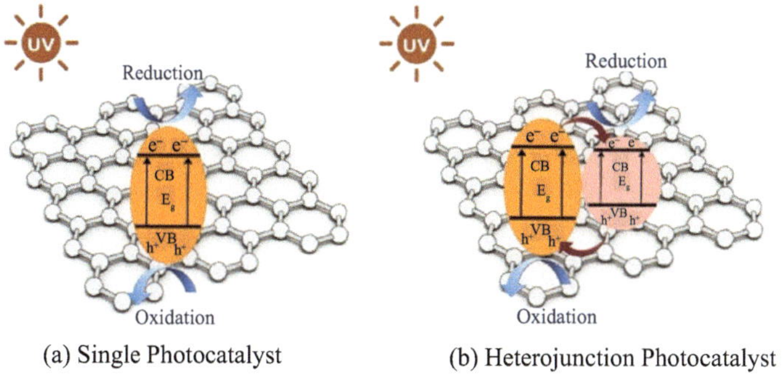


**Figure 3.1** Schematic illustration of (a) single photocatalyst and (b) heterojunction catalyst anchored on a Gr sheet.

The conversion to a three-dimensional (3D) Gr-based structure enables the retainment of the salient properties of pristine Gr. 3D Gr architectures which include hydrogel, aerogel, film, and foam have shown interesting performance in the preparation of high-performance remediators in environmental applications. The macroscopic 3D architectures are explored as a viable option as they effectively avoid rGO sheets from agglomerations, hence fabricating a porous framework exhibiting a large specific surface area. The incorporation of photoactive materials within the 3D rGO network results in the design of novel functional materials. Specifically, the unique properties of the 3D macrostructure allow for the anchoring of various photocatalysts which address the limitation brought by powder-based photocatalyst. Often, the poor reusability and regeneration of conventional photocatalyst are not attractive in real-time applications. To date, various synthetic protocols have been proposed for 3D Gr formation which includes hydrothermal and chemical reduction methods. The general idea of why 3D Gr is preferred in wastewater remediation is due to the convenient retrieval process in a solid-liquid interface. The advantages in terms of high reactivity and separation of free bulk enable the photocatalysts to be recycled after successive usage. In addition, 3D Gr is mechanically stable and able to withstand high compression which is a plus point in practical applications. Some studies have also included magnetic-based nanoparticles (NPs) on the 3D hydrogel which provides easier recovery of the hybrid photocatalysts [6, 7]. Numerous strategies for efficiently separating photogenerated electron-hole pairs in semiconductor photocatalysts have been proposed, including doping, metal loading, and/or the introduction of heterojunctions. The selection of various hybrid photocatalytic materials, especially heterojunction photocatalysts have gained popularity due to their higher performance than single photocatalyst materials. The engineering of heterojunction in photocatalyst can be regarded as one of the effective methods to prepare high-performing photocatalyst due to its feasibility in electron-hole pair separation [8].

Figure 3.1 summarized the typical photocatalytic process on a single photocatalyst and heterojunction catalyst. Briefly, a heterojunction photocatalyst is the combination of more than one semiconductor which are in contact with one another. The selection of visible light photocatalysts, with a smaller bandgap, is often desired due to their ability to perform photocatalysis under normal light conditions. Among other properties sought after are such as stability and biocompatibility. Due to its high photochemical stability and ability to efficiently utilize sunlight, heterogeneous photocatalysis is a feasible solution. Ideal photocatalysts should be non-toxic, easy to produce, possess high crystallinity, and not generate toxic by-products [9–11]. The photocatalytic reaction in a heterojunction photocatalytic system often follows the following step: (a) absorption of light, (b) formation of photogenerated electron-hole pairs, (c) migration of photogenerated electron-hole pair from one semiconductor to the other, (d) adsorption of pollutants, and (e) degradation by reactive oxygen species (ROS). In wastewater remediation by AOP, the commonly reported radicals to mineralize the organic compounds are $\cdot O_2^-$, $h^+$, and $\cdot OH$. In the case of hybrid 3D Gr, the Gr layers provide a web of electron percolation, hence proving electron migration. In addition, the Gr sheet improves photogenerated electrons and increases charge separation between the semiconductor. Often, the setback from 3D hybrid Gr is the excessive amount of rGO, which shields the light absorption ability, which is fundamental in the photocatalysis process. Moreover, the excess rGO can act as a recombination center, promoting the recombination of electron-hole pairs [12, 13]. Thus, an alternative strategy for optimizing the photocatalytic performance of these Gr-containing materials is to use a rational structural design of Gr sheets capable of mitigating their adverse effects. Various authors have systematically reviewed the recent trend in Gr-based materials applied for the decontamination of wastewater [14–16]. However, the usage of 3D macrostructure for photocatalysis usage has not been properly addressed. As such, this chapter's primary objective is to familiarize readers with hybrid 3D Gr with emphasis on their applications in wastewater remediation via photocatalysis. A brief overview of 3D Gr on its green preparation route will first be presented via chemical reduction and hydrothermal synthesis. The evaluation of their photocatalytic properties toward the remediation of synthetic dyes, heavy metals, and emerging contaminants are discussed, in addition to the mechanism of degradation by the ROS. The photocatalytic disinfection toward selected bacteria and their mechanism is also presented. Finally, we will discuss some future perspectives of hybrid 3D Gr for photocatalysis. Thus, it is hoped that this write-up will serve as a point of reference to a new entrance in the field intending to employ hybrid 3D Gr photocatalytic materials.

## 3.2 Self-Assembly of Hybrid 3D Gr Architecture

GO is often used as the starting material for the preparation of the 3D macrostructure ascribed to the ample oxygenated groups present which provide many active sites for interactions [17–19]. A combination of non-covalent interactions, for instance, hydrogen bonding and π-π interactions, is responsible for the self-assembly process. Hybrid 3D Gr-based photocatalyst preparation via the one-pot synthesis method involving the homogenous mixing of the GO with the reactants is often preferred due to their facile synthetic route as shown in Fig. 3.2. The photoactive materials are often introduced in the GO suspension before hydrothermal reaction to drive the interaction with the Gr sheet and also to form the photoactive materials. One of the leading synthetic protocols in the preparation of 3D Gr architecture is achieved via hydrothermal treatment. This thermal treatment method often employs hot-compressed water to drive the reaction which is often regarded as green, facile, and cost-effective [20]. The process is often initiated with the sealing of GO in a Teflon-lined vessel and subjected to various temperature elevations. The hydrogen ions produced from superheated water act as a reducing agent which exfoliates and decreases the surface-oxygenated functional groups. The reduction process drives the rearrangement of the graphitic aromatic structure, promoting π-π interactions and hydrophobic interactions [21]. The hydrothermal optimal condition is another factor to prepare various morphology of the photoactive materials as previously reported for improvement in their photocatalytic activity [22]. In some cases, ascorbic acid (AA) is added into the mixture to act as the reducing agent and also cross-linkers. Li et al. [23] hydrothermally prepared a 3D $TiO_2$/rGO in the presence of AA as the reducing agent and cross-linking agent.

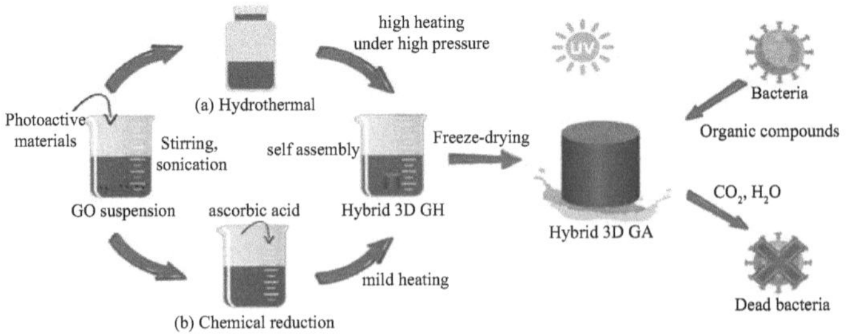


**Figure 3.2** Common methods for the preparation of hybrid 3D Gr photocatalyst. (a) hydrothermal method, (b) chemical reduction using AA for the remediation of organic compounds and pathogens.

In one study, the hydrothermal method was applied to prepare a Z-scheme 3D g-$C_3N_4$-ZnO/Gr heterojunctions for the adsorption and photocatalysis of various dye molecules [24]. To prepare the 3D hybrid material, g-$C_3N_4$ was first synthesized using heat decomposition from melamine. The ZnO was prepared using $Zn(NO_3)_2$ as the precursor. Both the photoactive materials were then added into the GO suspension, sonicated, and subjected to hydrothermal treatment at 180°C for 24 h. The reduction of GO using AA) was first reported by Zhang et al. [25]. Since then, many studies have reported on a 3D Gr formation using AA as the green reducing agent. Often AA is mixed into the GO suspension and mildly heated without agitation to drive the self-assembly to a Gr-monolith structure. This one-pot method allows for the preparation of various hybrid 3D Gr structures by merely mixing the photocatalytic materials in the GO suspension. The preparation of a TiO/rGH using commercially obtained Degussa $TiO_2$ (P25) as the photoactive material was previously reported [26]. In the synthetic protocol, an equal volume of PEG 200 and $TiO_2$ was added into GO and subjected to ultrasonication. Following that, sodium ascorbate was introduced as the self-assembly agent by heating the suspension at 95°C for 60 min to obtain the 3D hybrid material. Recently, the in-situ growth method was reported to prepare a hybrid 3D $BiVO_4$/rGH for the photodegradation of tetracycline (TC) [27]. The one-step method was initiated by adding $Bi(NO_3)_3.5H_2O$ and various mass fractions of reduced graphene hydrogel (rGH) in distilled water, followed by sonication and stirring. Then, $NH_4VO_4$ and $CO(NH_2)_2$ were added respectively and stirred. The suspension was heated at 80°C for 24 h to obtain the 3D photocatalytic material. They posited that the directional migration of $BiVO_4$ and rGH substantially inhibit photogenerated carriers from recombination, thus increasing the separation efficiency.

## 3.3 Altering the Properties of 3D Gr for Improved Photocatalytic Performance

The tuning of Gr structure is one important basis to enhance their photocatalytic ability. It is well known that Gr has excellent electron mobility which can assist in charge transferability. Three-dimensional (3D) hybridization of Gr sheets represents an extremely promising route to high-performance material to promote carrier transfer and stability. Fan et al. [28] recently reported on the blending of Gr and carbon nanotubes (CNTs) using the hydrothermal method for the creation of 3D neural-like hybrids to impart enhanced carrier transfer, structural stability, and electrocatalytic durability. The presence of abundant oxygen-containing groups in GO reacts with CNTs, forming $sp^3$-like defect structures in the folded interface region thus demonstrating improvement in their carrier transfer and stability. Hence, the proposed synthetic modification of Gr with other carbon-based materials is an ingenious strategy to enhance the photocatalysis process.

The doping of Gr structure by elements with varying electronegativities can significantly alter the electron mobility which is also an important aspect of photocatalysis. Often, doping with various dopants such as heteroatoms has been reported to significantly improve the performance of the hybrid system as a whole. Generally, the doping can be accomplished by a single element doping or via multielement doping. The simultaneous coupling of the co-dopants properties in the 3D template enables the establishment of superior attributes in comparison to a single dopant, hence enabling the ability for selective catalysis activity [29]. The improved photocatalytic ability from N introduction hinders the aggregation of the $BiVO_4$ and the Gr sheet [30]. Besides, the separation efficiency of the electron-hole pairs can be improved with the incorporation of the pyridine nitrogen, which successfully increases the photocatalytic degradation toward MB. Jiang et al. [31] also found that the co-doping of the 3D porous Gr aerogel (GA) with the lowest ratio of B to N improved photoactivity significantly as the hybrid material showed the highest Bisphenol A (BPA) removal efficiency ascribed to the presence of the bonding states in the lattice of the aerogel. The higher content of pyridinic N and lower B-N defect facilitate ROS production. Several strategies were proposed to enhance the photocatalytic efficiency of 3D Gr, effectively separating the photogenerated electron-hole pair in semiconductors. Among them, the preparation of heterojunction-based catalysts such as p-n heterojunctions and Z-scheme heterojunctions oriented on 3D Gr has been demonstrated to be the popular strategy for the photodegradation of various organic pollutants [8].

## 3.4 Recent Trends of Hybrid 3D Gr in Pollution Remediation via Photocatalysis

No doubt, adsorption is often associated with the remediation process using 3D Gr ascribed to their porous structure and textural properties. Previously, many review chapters have explained the adsorption principle and its performance in the adsorption of various noxious pollutants [15,32,33]. In the case of hybrid 3D Gr, it was mentioned that the synergistic adsorption and photocatalysis processes were responsible for the remediation of various pollutants. Substantial studies have shown the reliability of the 3D Gr structure for the adsorption of various pollutants [34,35]. As previously discussed, 3D Gr significantly improved reactant accessibility and adsorption. In addition, improved utilization of visible light is due to the 3D porous structure, as opposed to their 2D derivatives. In this section, the authors will attempt to discuss the application of the various 3D hybrid Gr macrostructure for the photodegradation of pollutants with a specific focus on the usage of hydrothermally prepared Gr-based hydrogels. Several reviews have systematically addressed the application of Gr composite for the remediation of various organic contaminants [36–38].

The photodegradation of synthetic dyes has become the commonly used model pollutant to access the photodegradation performance of the hybrid 3D Gr. Synthetic dyes, which are very soluble in water, are often applied in the textile, paint, and printing industries. This has led to a systemic problem in their removal from wastewater. Briefly, the dyes used can be categorized into cationic dyes and anionic dyes. The chromophores and auxochromes are responsible for the color and affinity toward fibers, respectively. Wu et al. [39] prepared a hybrid GA anchored with molybdenum trioxide via the chemical reduction using AA and freeze-drying technique for the photodegradation of MB. They attributed the improved photodegradation efficiency from the improved electron migration and carrier lifetime and inhibited electron-hole combination to the 3D structure. The hydroxyl radical and radical anions were mostly responsible for the degradation of the MB.

The S-atom sharing strategy for the chemical bonding between $CoS_2$ and $MoS_2$ anchored on a 3D N-doped GA for the efficient photodegradation of Rhodamine B (RhB) has also been proposed [40]. $MoS_2$ can expose more active sites due to the sulfur-atom sharing strategy, and $CoS_2$ and N-GA expedited surface catalytic reactions. The N-GA enables the active sites to run continuously so that the accumulation process as the rate-determining phase will be avoided. In the photodegradation mechanism, when the sample is illuminated with visible light, $MoS_2$ nanosheets absorb photons, resulting in electron-hole pairs formation. The S-atoms bridge a connection between $CoS_2$ and $MoS_2$ which formed electron clouds for easier electron migration. Consequently, photoelectrons can be easily transferred to the N-GA to prevent electron-hole recombination [46]. The photoelectrons react with $O_2$ on the Gr sheet to form $\cdot O_2$, whereas the remaining holes can either degrade RhB directly or react with ·OH to form OH.

Emerging contaminants are derived from medical, pesticides, pharmaceutical, industrial chemicals, and personal care products which have become ubiquitous pollutants. These contaminants are often not removed completely during the wastewater treatment process attributed to their microsize/nanosize structure. Their presence in wastewater has raised concerns about their harmful effect on human health. A bismuth-based semiconductor has also garnered multiple attention for its feasibility in photocatalysis applications. In a recent study, a $Bi_2O_3/BiVO_4$ composite self-assembled on a GO network via vdW interactions was reported for enhanced oxygen generation [41]. The hydrothermal preparation of the photocatalysts enables the fabrication of mixed-dimensional vdW heterostructure. Their prepared heterostructure showed high photoactivity oxygen evolution, three times higher than pure $BiVO_4$ ascribed to the accelerated charge separation and transfer processes. Recently, $BiPO_4$/GA was synthesized via the one-step hydrothermal treatment for the mineralization of BPA and MB [42]. It was found that degradation of BPA and MB was 50.67% and 96.23%, respectively in which the main active species responsible for the degradation were superoxides and holes. Ma et al. [27] prepared a novel 3D $BiVO_4$/rGH/QDs self-decorated photocatalyst for the TC hydrochloride (TCl) degradation. The 3D hybrid structure provided extra photoinduced charge carriers and active sites, leading to enhanced photocatalytic performance. The hybrid 3D photocatalyst was compared to that of bare $BiVO_4$, which has a 2.2- and 3.7-

fold higher degradation efficiency and rate constant, respectively. The directional migration of $BiVO_4$ and rGH significantly inhibit photogenerated carriers from recombination, thereby improve the efficiency of their separation. $\cdot O_2^-$ and $h^+$ were found to be the main reactive species for the degradation of TCl.

The structural facet of the semiconductor is another important factor to take into consideration to improve their photocatalysis performance [43,44]. In one study, the synthesis of catalytic active {0 0 1} facets of $TiO_2$ orientated on Gr foam was proposed for enhanced photoactivity [45]. The preparation of the Gr foam proceeded via vacuum filtration of GO and the gasification by hydrazine hydrate. Following that, solvothermal treatment at 180°C for 12 h of the Gr foam in $TiF_4$ and diethylene glycol to fabricate the high energy active facet $TiO_2$. The synergistic effects of intrinsic $TiO_2$ with highly exposed {0 0 1} facets and the strong bond between $TiO_2$ and Gr contributed to the remarkable photocatalytic ability of $TiO_2$/Gr foam. The abundant active sites from the facets drive surface catalytic reactions for oxygen activation and interfacial in-plane and vertical charge transfer.

High-performing heterogeneous catalysts for the generation of free radicals using persulfate activation are gaining traction in wastewater application for their environmentally friendly strategy. Ample metal oxide and metal sulfide materials have been previously explored to generate $SO_4^{\cdot-}$, which is from the activation of peroxymonosulfate (PMS) to degrade pollutants [46]. Recently, Zou et al. [47] proposed a porous $CoFeO_4$/rGO (CFGA) aerogel via the hydrothermal and freeze-dried for the degradation of phenol. Their CFGA enabled efficient activation of PMS and rapid phenol degradation. The salient properties of rGO aerogel of possessing a large surface area and 3D porous structure further facilitate the mass transfer of the $CoFe_2O_4$ catalytic sites. Consequently, the outstanding PMS-activation performance of CFGA was evidenced by the quick removal of phenol over a wide pH range in the presence of chloride or humic acid. Mechanistic studies indicated that the Co atom mediates the PMS activation via a radical pathway in the CFGA. In a recent study, He et al. [48] found that the 3D GA provided an efficient electron transfer pathway for the degradation of metronidazole via PMS activation. The $Co_3Mn$-LDH/rGA/PMS/Vis system offered a promising avenue to continuously degrade various organic pollutants. Yi et al. prepared a 3D cobalt hydroxide/GO hydrogel for automated laboratory-scale wastewater treatment via PMS activation [49]. They found that the $Co^{2+}$ incorporated within the 3D Gr framework allows for the generation of active species. The GO and the polyacrylic acid (PAA) hydrogel functioned to improve the catalyst lifetime and also reduce secondary pollution from the cobalt ion.

The preparation of a 3D $ZnSnO_3$/Gr via hydrothermal and freeze-drying for the photodegradation of ciprofloxacin was recently reported [50]. The improved photoactivity of the hybrid 3D macrostructure is attributed to the reduced bandgap structure of the $ZnSnO_3$, possessing a special crystal structure to reduce the recombination of photo degenerated electron holes leading to greater ·OH, and $\cdot O_2^-$ active species. In addition, the 3D Gr structure possessing a high specific surface area provided additional reaction sites to improve the photocatalysis process. In another study, Rezaei et al. [51] prepared a 3D hybrid rGH-$Fe_3O_4$/$SnO_2$/Ag using AA as the reducing agent for the synergistic adsorption and photocatalysis of 2,4 dichlorophenol (DCP). They first prepared the photoactive material composed of $Fe_3O_4$/$SnO_2$ as the cores and Ag as the coating. Following that, AA was used to drive the self-assembly into the 3D structure. The adsorption capacity of the DCP using 10% rGH-$Fe_3O_4$/$SnO_2$/Ag was 14.90 mg/g after 0.5 h. Increased $Fe_3O_4$/$SnO_2$/Ag content appeared to reduce adsorption capacity while increasing photocatalytic performance rate. The degradation efficiency was recorded at 93.8% under visible light irradiation. They found that photo-generated holes played the most significant role in the oxidation of DCP after examining the photocatalytic mechanism via electron spin resonance spectrometer (ESR) and radical trapping experiments. In the proposed mechanism, the photogenerated holes directly reacted with the DCP and water to form ·OH. The rapid charge transfer facilitated by rGH enables the reaction with $O_2$ to produce $\cdot O_2^-$. The irradiation on the photoactive 3D material leads to the generation of e- which are transferred from $SnO_2$ to Ag and, to the rGH surface to further react with $O_2$ to produce $\cdot O_2^-$. In addition, the $h^+$ can readily degrade DCP. The proposed mechanism of degradation is illustrated in Fig. 3.3.

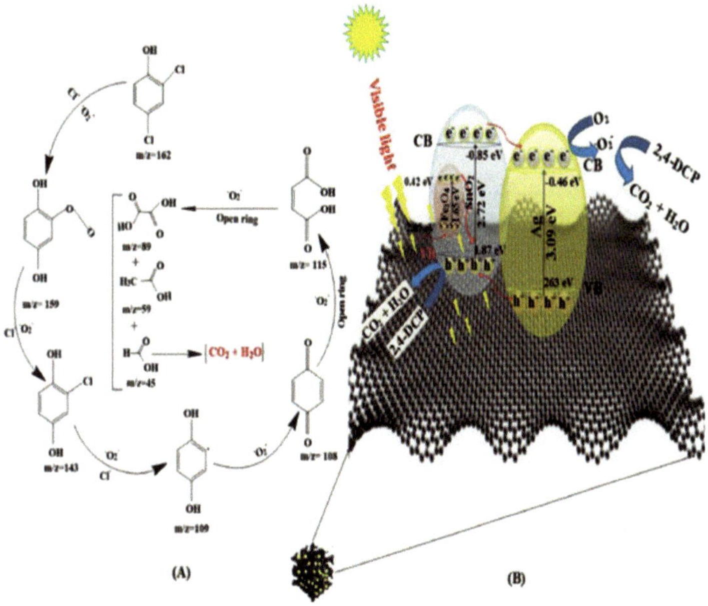


**Figure 3.3** Proposed degradation pathway of DCP using 10% rGH-$Fe_3O_4$/$SnO_2$/Ag and the suggested photodegradation mechanism under visible sunlight. (Reprinted with permission from Rezaei et al., *J. Taiwan Inst. Chem. Eng.* 2021, Elsevier.)

The inclusion of g-$C_3N_4$ in the preparation of novel heterojunction photocatalyst material can be attributed to their ability to be tuned in terms of their microstructure for active sites creation and enlargement of surface area for light absorption [52]. Recently, the preparation of an S-scheme CdS-g-$C_3N_4$-GA was reported [53]. The synthetic protocol to prepare the hybrid 3D is via hydrothermal and ultrasonic assistance. The ternary heterojunction exhibited excellent photocatalytic activity from the effective carrier separation and porous framework of the 3D Gr. The photodegradation efficiency of RhB reached up to 91.4% within 50 min for CdS-g-$C_3N_4$-GA, which was approximately 40 and 11 times better than that of g-$C_3N_4$ and CdS, respectively. Due to the strong internal electric field used to accelerate photo-induced carriers separation, the superior photocatalytic performance of S-scheme heterojunction makes it an ideal candidate for the design of high-performance photocatalytic systems. In another study, Guo et al. [54] fabricated a 3D hybrid structure composed of 3D rGO/g-$C_3N_4$/NiO/$Ni_3(BO_3)_2$

by embedding flower-like $NiO/Ni_3(BO_3)_2$ microspheres to form a Z-scheme system with g-C3N4. The hybrid material outperformed g-$C_3N_4$ and $NiO/Ni_3(BO_3)_2$ in photocatalytic degradation of MB and TC. Their result demonstrated that rGO aerogel simultaneously improved g-$C_3N_4$ dispersibility and charge transfer efficiency. The dual Z-scheme heterojunction photocatalyst possessed wide light absorption and also high redox efficiency toward MB and TC. Additionally, when compared to the binary structure, the novel hybrid material significantly improves charge transfer efficiency and retards electron-hole recombination. The preparation of a ternary composite based on 3D rGO wrapped with nanoscale zero-valent iron (nZVI) doped with $Al_2O_3$ for the generation of ROS via the activation of dissolved oxygen was also reported. The photodegradation of chloramphenicol via Fenton-like degradation was achieved from the generation of $^1O_2$, ·OH, and active sites of $Fe^{2+}$ [55].

3D hybrid Gr material was also reported for the remediation of a heavy metal ion such as for Cr (VI) photoreduction to Cr (III). Cr (VI) is water soluble and extremely toxic, while Cr (III) is less soluble and precipitates in the absence of complexing ligands. Thus, under controlled conditions, the easy reduction of Cr (VI) to Cr (III) is desired. Cr (VI) is a commonly used heavy metal pollutant in water and is produced by printing, electroplating, dyeing, manufacturing, and the tanning process. The photoreduction of the metal ion is typically comprised of three important steps. The first step is often the adsorption of the Cr (VI) on the 3D Gr surface. As mentioned earlier, the porous structure of the 3D Gr allows for physical or chemical adsorption to occur. After irradiation with UV/Vis light, electrons are excited, and holes are formed. The electrons migrated to the semiconductor surface for the reduction of Cr (VI) to Cr (III) or $H^+$ to $H_2$. Concurrently, the hydroxyl and carboxylic groups of the rGO consumed the photogenerated holes. Briefly, the reduction of Cr (VI) often follows the following equations.

$$\text{(Hybrid 3D Gr photocatalysts)} + h\nu \rightarrow h^+ \text{ (valence band)} + e^- \text{ (CB)} \quad (3.1)$$

$$H_2O + 4h^+ \rightarrow O_2 + 4H^+ \quad (3.2)$$

$$Cr_2O_7^{2-} + 14H^+ + 6e^- \rightarrow 2Cr^{3+} + 7H_2O \quad (3.3)$$

In one study, Zhang et al. [56] prepared a visible light response

3D $MoS_2$/rGO aerogel via the hydrothermal and freeze-drying and accessed their photoreduction toward Cr (VI). After 2 h of photoreduction, the $MoS_2$/rGO aerogel reduced 92% of the Cr (VI) to Cr (III), whereas the $MoS_2$ powder demonstrated almost no photoreduction activity for Cr (VI). They discovered that the 3D GA architecture inhibited nanoparticle aggregation and enabled the composite to form interconnected porous architectures with excellent electrical conductivity. The hybrid GA enabled charge carrier separation improvement and accelerated transfer of photoelectrons to the $MoS_2$ surface.

## 3.5 Photocatalytic Disinfection Using Hybrid 3D Gr

In recent years, significant attention has been paid to the disinfection photocatalytic properties of hybrid 3D Gr materials in aqueous media. Microorganism contamination can impair the photocatalytic efficiency of the catalyst by biofilm formation which reduces the light intake, and thus, materials with antibacterial properties can significantly address this inherent issue. Even without functional modification, Gr-based materials have emerged as antimicrobial agents [17]. These Gr derivatives are capable of effectively inhibiting *E. coli* growth while exhibiting low cytotoxicity. Often the treatment of pathogens in wastewater via photocatalysis is associated with the generation of ROS such as $\cdot O_2^-$, $^1O_2$, $\cdot OH$, and $H_2O_2$ from the oxidation and reduction reactions. It has been demonstrated that ROS is capable of inducing cell death in a wide variety of Gram-negative and Gram-positive bacteria. The cell wall and cytoplasmic membrane are degraded as a result of the production of ROS such as hydroxyl radicals and hydrogen peroxide. This results in the leakage of cellular contents, followed by cell lysis and, in some cases, complete mineralization of the organism. The technology has the potential to be a highly effective weapon in the fight against various pathogens in wastewater, especially with the development of visible light-activated catalysts. Singh et al. [57] have systematically reviewed the usage of Gr-based material for bacterial disinfection.

**Table 3.1** Recently reported hybrid 3D graphene used in the photocatalysis of various pollutants which include synthetic dyes, emerging contaminants, and heavy metals

| Hybrid 3D composite | Preparation method | Light source | Model pollutant | Main species | Photocatalytic efficiency |
|---|---|---|---|---|---|
| $rGO/g\text{-}C_3N_4/NiO/Ni_3(BO_3)_2$ (GACN/NNB) [54] | HT and freeze-drying | Vis, 300 W, Xe | MB (20 mg/L, 50 ml)<br>TC (10 mg/L, 50 ml) | $\cdot OH$ and $\cdot O_2^-$ | 96.9% (60 min), MB<br>94.0% (30 min), TC |
| $CoS_2/MoS_2$-N-GA [40] | HT and vacuum drying | Vis, 300 W, Xe | RhB (10 mg/L, 50 ml) | $\cdot O_2^-$ and $\cdot OH$ | 90% (40 min) |
| $MoO_3$/rGO [39] | Chemical reduction using ascorbic acid and freeze-drying | Vis, 300 W, Xe | MB, MO (40 mg/L, 200 ml) | $\cdot OH$ and $\cdot O_2^-$ | 90% (120 min), MB<br>16.8% (120 min), MO |
| $Bi_7O_9I_3$/N-GA [58] | HT and freeze-drying | Vis, 200 W, Xe | RhB (10 mg/L, 150 ml) | $\cdot O_2^-$ and $h^+$ | 96.3% (100 min) |
| $Ag/g\text{-}C_3N_4$/rGO [59] | HT and freeze-drying | Vis, 300 W, Xe | R6G (10 ppm, 35 ml) | $\cdot O_2^-$, $h^+$, and $\cdot OH$ | 93.0% (240 min) |
| Tannic acid-$TiO_2$/rGO [60] | HT and freeze-drying | Vis, 300 W, Xe | RhB, MB, TC (10 mg/L, 50 ml) | $\cdot O_2^-$ and $\cdot OH$ | 91.9% (180 min), RhB<br>77.8% (180 min), MB<br>43.2% (180 min), TC |
| GA/$Bi_2WO_6$ [61] | HT and freeze-drying | Vis, 350 W, Xe | RhB (10 and 15 mg/L, 200 ml)<br>SN (10 mg/L, 200 ml) | $\cdot O_2^-$, $h^+$, and $\cdot OH$ | 97.49% (120 min), RhB<br>55.81% (120 min), SN |
| GA-CQDs/CNN [62] | HT and freeze-drying | Vis, 300 W, Xe | MO (30 mg/L, 100 ml) | $\cdot O_2^-$ and $\cdot OH$ | 91.1% (240 min) |

| Hybrid 3D composite | Preparation method | Light source | Model pollutant | Main species | Photocatalytic efficiency |
|---|---|---|---|---|---|
| ZnO/GA [63] | HT and freeze-drying | Vis, 300 W, Xe | MB (20 mg/L, 25 ml) | $\cdot O_2^-$ and $\cdot OH$ | 97.6% (180 min) |
| Ag-AgBr/ $BiVO_4$/GA [64] | HT, UV irradiation, and freeze-drying | Vis, 500 W, Xe | MO (10 mg/L, 20 ml) | $\cdot O_2^-$ and $h^+$ | 93.92% (24 min) |
| g-$C_3N_4$/α-Fe2$O_3$/GA [65] | HT and freeze-drying | Vis, 300 W, Xe | MB ($3\times10^{-5}$ M, 30 ml) | $\cdot O_2^-$, $h^+$, and $\cdot OH$ | 76.5% (180 min) |
| GA/$Sb_2WO_6$ [66] | Electrostatic self-assembly, HT and freeze-drying | UV, 11 W UV and Vis, 300 W, Xe | MO (6 mg/L, 500 ml) | $\cdot O_2^-$ and $h^+$ | 57.8% (300 min) |
| g-$C_3N_4$-ZnO/GA [24] | HT and freeze-drying | UV, 300W Hg and Vis, 300 W, Xe | RhB (20 mg/L, 25 ml) | $\cdot O_2^-$ and $\cdot OH$ | 81.0% (120 min, Vis)<br>82.7% (120 min, UV) |
| CuS-NS/rGO [67] | Chemical reduction using ascorbic acid and freeze-drying | Vis, 500 W, Xe | RhB, MB (10 mg/L, 60 ml)<br>Cr (VI) (20 mg/L, 60 ml) | $\cdot O_2^-$, $h^+$, $\cdot OH$, e- | 99.3% (60 min, RhB)<br>99.15% (60 min, MB)<br>98.9% (80 min) |
| NGA-TEPA [68] | HT and freeze-drying | Vis, 11 W, Xe | AO ($7.5 \times 10-5$ M, 18 ml) | $\cdot O_2^-$ and $\cdot OH$ | 95% (300 min, MB) |
| $Bi_2S_3$/$BiVO_4$ GA [69] | HT and freeze-drying | Vis, 300 W, Xe | Cr (IV) (50 mg/L, 50 ml)<br>BPA (10 mg/L, 50 ml) | $e^-$<br>$\cdot O_2^-$, h+, and $\cdot OH$ | 100% (40 min, adsorption + 120 min photocatalysis) |

(*Continued*)

**Table 3.1** (*Continued*)

| Hybrid 3D composite | Preparation method | Light source | Model pollutant | Main species | Photocatalytic efficiency |
|---|---|---|---|---|---|
| $Zn_{0.5}Cd_{0.5}S$/rGO [70] | HT and freeze-drying | Vis, 500 W, Xe | Cr (IV)(20 mg/L, 50 ml) | $e^-$ | 95% (140 min) |
| $TiO_2$/rGH [26] | Chemical reduction using sodium ascorbate and vacuum drying | UV, 250 W, Hg | Cr (IV) (5 mg/L, 100 ml) | $e^-$ | 100% (30 min) |
| AgBr/B-rGO [71] | HT and freeze-drying | Vis, 300 W, Xe | Cr (IV) (15 mg/L, 20 ml) | $e^-$ | 95.80% (60 min) |
| $MoS_2$/rGO [56] | HT and freeze-drying | Vis, 300 W, Xe | Cr (IV) (100 mg/L, 150 ml) | $e^-$ | 92% (120 min) |
| $BiVO_4$/rGH [27] | In situ growth method and oven-drying | Vis, 300 W, Xe | TC (20 mg/L, 100 ml) | $\cdot O_2^-$, $h^+$ | 73.2% (120 min) |
| $TiO_2$-GA [72] | HT and freeze-drying | UV, 24 W UV lamp<br>Vis, 1.2 W VUV lamp | EZ (100 ppb, 150 ml) | $h^+$, and$\cdot$OH, $\cdot O_2^-$ | 98.5% (60 min)<br>99.8% (3 min) |

*Abbreviations*: AO, acridine orange; BPA, Bisphenol A; EZ, ethenzamide; HT, hydrothermal; SN, sulfanilamide; TC, tetracycline.

Silver-based semiconductors such as $AgVO_3$, AgBr, $Ag_2S$, and $Ag_3PO_4$ have been extensively reported to not only possess antibacterial properties but also induce separation of electron-hole pairs efficiency upon light irradiation. The fabrication of a 3D visible-light-activated Ag–AgBr/$TiO_2$/GA composite for enhanced photocatalytic degradation of *E. coli* and *S. aureus* was recently reported [73]. After irradiation for 42 min, AgAgBr/$TiO_2$/GA exhibits superior photocatalytic inactivation with the amount of viable bacteria for *E. coli* and *S. aureus* is 0 and $10^{3.2}$, respectively, suggesting that the heterojunction system formed by $TiO_2$ and the $Ag_0$/AgBr enhanced antibacterial efficiency. In another study, Chen et al. [74] prepared a 3D porous $Ag_2S$/$AgVO_3$/GA using the in-situ ion exchange method and hydrothermal treatment for photocatalytic antifouling experiments. Two model bacterial pollutants, *E. coli* and *S. aureus* were applied to access the photocatalytic ability of the hybrid 3D material. The inactivation efficiency of *E. coli* is higher than that of *S. aureus* which is due to the thicker cell wall of the Gram-positive bacteria. Further, the antifouling mechanism was accessed using the radical trapping test which showed that the $\cdot O_2^-$ and $h^+$ radicals were the main species responsible for the *E. coli* inactivation. The antibacterial cycle test results for *E. coli* indicated that the 3D hybrid Gr maintained after five consecutive cycles. Hence, it is believed that the template provided by the 3D Gr allows for the repeated usage of the photocatalyst even in antifouling applications. The preparation of a 3D AgBr/Ag/Gr via chemical reduction, immersion in cetrimonium bromide (CTAB) and freeze-drying was reported [75]. The composite is more successful than pure AgBr at inhibiting bacteria during photocatalysis. It has been demonstrated that electron relays formation provides a significant effect on the photogenerated charge carriers separation resulting in increased photoactivity for the 3D hybrid Gr material. The "electron sink" effect of the Ag NPs during photocatalytic disinfection results in increased charge transfer and photoactivity. Additionally, the hot electrons of Ag NPs excited by SPR are beneficial for increasing photoactivity. Additionally, the composite demonstrates superior recyclability and operability in recycling tests. Lin et al. [64] prepared Ag-AgBr/$BiVO_4$/GA using the hydrothermal method and freeze drying, accessed their antibacterial activity under dark and light condition. In dark condition, their hybrid 3D Gr is not toxic to the *E. coli* and

*S. aureus*. However, in light condition for 24 min, the activity reached 7 log and 2.9 log for *E. coli* and *S. aureus*, respectively as shown in Fig. 3.4(a,b). Furthermore the hybrid 3D Z-scheme photocatalyst also exhibited excellent antibacterial activity after three cycles which is portrayed in Fig. 3.4(c). The findings further showed that heterojunction photocatalysts combined with Gr possessed practical value in wastewater remediation applications.

Copper-based nanomaterials are another common antibacterial material that has seen its vast usage in wastewater remediation applications [76]. In one study, Ma et al. [77] accessed the antibacterial disinfection of CuI-BiOI/rGO hydrogel under visible light. They prepared the hybrid 3D Gr using the solvothermal method and found that their hybrid 3D Gr was able to inactivate *E. coli* and *S. aureus* under visible light irradiation for 40 min. Under visible light irradiation conditions, the 3D hybrid Gr showed improved antibacterial activity in comparison to CuI-BiOI powder and BiOI/rGO hydrogel. The fluorescence staining method using DAPI and PI under various reaction time was used to depict the morphological changes of *E.coli*. It was discovered that the fluorescence produced by the two techniques was nearly identical, suggesting that *E. coli* had been entirely inactivated by the process. As for the *S. aureus*, the hybrid 3D Gr also exhibited excellent sterilization performance. Hence, they posited that the CuI and Gr promoted the photogenerated carrier separation. The ·OH and $H_2O_2$ with strong oxidation ability inherently damaged the bacterial physiological structure leading to cell death.

Sulfate radical-based AOPs generated from two major oxidants, peroxydisulfate and PMS are another important technique for antibacterial applications. The synthesis of 3D rGO foams for enhanced antibacterial activity toward *E. coli* via PMS photocatalysis by hydrothermal and solvothermal methods from GO was reported [78]. The findings indicated that in the absence of PMS, the hydrothermally prepared 3D rGO photocatalytic inactivation proceeded via a mixed radical and non-radical pathway, as evidenced by the moderate inhibition of bacterial inactivation by the used radical scavengers. The researchers proposed that the •OH and $SO_4^-$ were generated from the visible light irradiation which resulted in the $e^-$ generation and PMS generation. Following that, the generation of the oxidative $SO_5^{\cdot -}$ which is most likely capable of inactivating bacteria can be observed as a result of the sulfate and hydroxyl radicals' interchangeability.

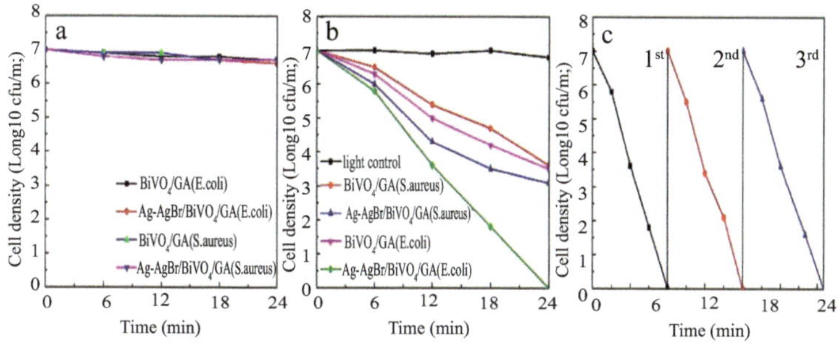


**Figure 3.4** Inactivation of *E. coli* and *S. aureus* ($10^7$ CFU/ml) by Ag-AgBr/$BiVO_4$/GA in (a) dark and (b) under visible light (>420 nm). (c) Recycling test for photocatalytic disinfection (Reprinted with permission from Lin et al., *Colloids Surfaces A Physicochem. Eng. Asp.* 2020, Elsevier).

## 3.6 Conclusion and Future Perspectives

Without a doubt, hybrid 3D GA has a promising future due to its unique characteristics and features that can provide ample research and development opportunities that are not exclusive only to water-remediation applications. The preparation of hybrid 3D GA via the hydrothermal treatment has shown ample opportunities for the design of novel photocatalyst for the remediation of various noxious pollutants. The hydrothermal treatment is a feasible yet simple approach for the immobilization of various heterojunction photocatalysts on the 3D Gr architecture. The addition of more than one co-catalysts has become the common strategy to make full use of the merits of the heterojunction-based photocatalyst. In this case, the 3D Gr framework enables the improvement of the light absorption ability, charge transport, and reduced recombination. The chemical interaction between the catalysts such as sharing of electrons plays a pivotal role in further improving the photoactivity of the hybrid 3D Gr. The determination of the optimal weight percentage of Gr needed to enhance the catalytic activity is a critical factor in enhancing the composite's activity. As expected, enhanced photocatalytic bacterial inactivation performance of hybrid 3D Gr derived from Ag and

Cu-based semiconductors were observed which bring an added value for application in real-time conditions. Even though many metallic semiconductors have been incorporated with 3D Gr, non-metallic catalysts should be taken into consideration as some have contributed to enhanced photocatalytic remediation. For example, g-$C_3N_4$ is one of the most studied photocatalytic materials; however, a rational combination of 3D Gr and g-$C_3N_4$ for photocatalytic degradation can be a viable substitute. The issue of water pollutants will continue to affect humankind amidst industrialization and modernization. It is expected that the production and release of contaminants will eventually be able to circumvent conventional monitoring technologies. As for the application of hybrid 3D Gr in practical applications, it is paramount to address several issues: (i) stability of the novel material, (ii) ability to remediate various contaminants at one time, (iii) excellent reusability and regeneration. The field of hybrid 3D Gr photocatalyst is ever-growing, which requires systematic and comprehensive research. To summarize, we believe that hybrid 3D Gr photocatalyst is a promising avenue for mass commercialization from its current potential in photocatalysis application.

## Acknowledgments

The first author would like to acknowledge the Ministry of Education, Culture, Sports, Science, and Technology, Japan for the scholarship awarded for his PhD studies. We would like to also thank Mr. Tithiphong Sukheeket for the visualization artwork.

## Conflict of Interest

The authors declare that they have no known competing financial interests or personal relationships that could have appeared to influence the work reported in this chapter.

## References

1. V. Singh, D. Joung, L. Zhai, S. Das, S. I. Khondaker, and S. Seal, *Prog. Mater. Sci.*, 2011, **56**, 1178.
2. H. Wang, T. Maiyalagan, and X. Wang, *ACS Catal.* 2012, **2**, 781.

3. P. Avouris and C. Dimitrakopoulos, *Mater. Today* 2012, **15**, 86.
4. H. Lu, S. Zhang, L. Guo, and W. Li, *RSC Adv.* 2017, **7**, 51008.
5. A. L. T. Zheng, S. Boonyuen, G. Y. Li, L. H. Ngee, and Y. Andou, *J. Mol. Struct.* 2021, **1245**, 131008.
6. C. Santhosh, A. Malathi, E. Daneshvar, P. Kollu, and A. Bhatnagar, *Sci. Rep.* 2018, **8**, 1.
7. A. Ziarati Saravani, M. Nadimi, M. A. Aroon, and A. Ebrahimian Pirbazari, *J. Alloys Compd.* 2019, **803**, 291.
8. J. Low, J. Yu, M. Jaroniec, S. Wageh, and A. A. Al-Ghamdi, *Adv. Mater.* 2017, **29**, 1601694.
9. B. Ohtani, *J. Photochem. Photobiol. C Photochem. Rev.* 2010, **11**, 157.
10. J. Sun, M. Zhang, Z. -F. Wang, H. -Y. Chen, Y. Chen, N. Murakami, and T. Ohno, *Rare Met.* 2019, **38**, 287.
11. S. W. Verbruggen, *J. Photochem. Photobiol. C Photochem. Rev.* 2015, **24**, 64.
12. G. Žerjav, M. S. Arshad, P. Djinović, I. Junkar, J. Kovač, J. Zavašnik, and A. Pintar, *Nanoscale* 2017, **9**, 4578.
13. C. Xu, X. He, C. Wang, X. Chen, R. Yuan, and W. Dai, *RSC Adv.* 2016, **6**, 84068.
14. C. Foti, P. G. Mineo, A. Nicosia, A. Scala, G. Neri, and A. Piperno, *Front. Chem.* 2020, **8**, 608236.
15. F. Perreault, A. Fonseca de Faria, and M. Elimelech, *Chem. Soc. Rev.* 2015, **44**, 5861.
16. A. L. T. Zheng and Y. Andou, *Int. J. Environ. Sci. Technol.* 2021.
17. A. L. T. Zheng, S. Boonyuen, T. Ohno, and Y. Andou, *J. Porous Mater.* 2021, **28**, 1291–1300.
18. A. L. T. Zheng, S. Boonyuen, T. Ohno, and Y. Andou, *Processes* 2021, **9**, 169.
19. A. L. T. Zheng, T. Phromsatit, S. Boonyuen, and Y. Andou, *FlatChem* 2020, **23**, 100174.
20. I. Ibrahim, T. Tsubota, M. A. Hassan, and Y. Andou, *Processes* 2021, **9**, 149.
21. Y. Zhou, Q. Bao, L. A. L. Tang, Y. Zhong, and K. P. Loh, *Chem. Mater.* 2009, **21**, 2950.
22. J. Li, Y. Wang, H. Ling, Y. Qiu, J. Lou, X. Hou, S. P. Bag, J. Wang, H. Wu, and G. Chai, *Nanomaterials* 2019, **9**, 65.
23. Y. Li, J. Yang, S. Zheng, W. Zeng, N. Zhao, and M. Shen, *Ceram. Int.* 2016,

**42**, 19091.

24. J. -Y. Zhang, J. -Y. Mei, S.-S. Yi, and X.-X. Guan, *Appl. Surf. Sci.* 2019, **492**, 808.
25. J. Zhang, H. Yang, G. Shen, P. Cheng, J. Zhang, and S. Guo, *Chem. Commun.* 2010, **46**, 1112.
26. Y. Li, W. Cui, L. Liu, R. Zong, W. Yao, Y. Liang, and Y. Zhu, *Appl. Catal. B Environ.* 2016, **199**, 412.
27. C. Ma, W. C. Seo, J. Lee, Y. Kim, H. Jung, and W. Yang, *Chemosphere* 2021, **275**, 130052.
28. S. -S. Fan, L. Shen, Y. Dong, G. Tian, S. -M. Wu, G. -G. Chang, C. Janiak, P. Wei, J. -S. Wu, and X. -Y. Yang, *J. Energy Chem.* 2021, **57**, 189.
29. S. Ullah, M. Hasan, H. Q. Ta, L. Zhao, Q. Shi, L. Fu, J. Choi, R. Yang, Z. Liu, and M. H. Rümmeli, *Adv. Funct. Mater.* 2019, **29**, 1904457.
30. X. L. Wang, J. Li, and W. M. Liu, *Opt. Mater.* 2021, **114**, 110922.
31. Y. Jiang, S. Chowdhury, and R. Balasubramanian, *J. Environ. Chem. Eng.* 2020, **8**, 104300.
32. K. Yang, J. Wang, X. Chen, Q. Zhao, A. Ghaffar, and B. Chen, *Environ. Sci. Nano* 2018, **5**, 1264.
33. V. K. Gupta and Suhas, J. Environ. Manage. 2009, **90**, 2313.
34. F. Zhang, Y. -H. Li, J. -Y. Li, Z. -R. Tang, and Y. -J. Xu, *Environ. Pollut.* 2019, **253**, 365.
35. Z. Bano, S. A. Mazari, R. M. Y. Saeed, M. A. Majeed, M. Xia, A. Q. Memon, R. Abro, and F. Wang, *J. Water Process Eng.* 2020, **36**, 101404.
36. A. V. Karim and A. Selvaraj, *Process Saf. Environ. Prot.* 2021, **146**, 136.
37. O. C. Olatunde and D. C. Onwudiwe, *Int. J. Environ. Res. Public Health* 2021, **18**, 1529.
38. F. Khan, M. S. Khan, S. Kamal, M. Arshad, S. I. Ahmad, and S. A. A. Nami, *J. Mater. Chem. C* 2020, **8**, 15940.
39. T. Wu, B. Zhang, Z. Wu, J. Zhang, H. Liu, S. Yu, Z. Huang, and X. Cai, *RSC Adv.* 2019, **9**, 37573.
40. L. Zhang, L. Wu, Z. Feng, Q. Meng, Y. Li, and T. Duan, *J. Environ. Chem. Eng.* 2021, **9**, 104771.
41. Y. Bi, Y. Yang, X. -L. Shi, L. Feng, X. Hou, X. Ye, L. Zhang, G. Suo, J. Chen, and Z. -G. Chen, *J. Colloid Interface Sci.* 2021, **593**, 196.
42. W. Yang, S. Tang, Z. Wei, X. Chen, C. Ma, J. Duan, and R. Tan, *Chem. Eng. J.* 2021, 129720, **421**.
43. M. Kowalkińska, S. Dudziak, J. Karczewski, J. Ryl, G. Trykowski, and A.

Zielińska-Jurek, *Chem. Eng. J.* 2021, **404**, 126493.

44. T. Joutsuka, H. Yoshinari, and S. Yamauchi, Bull. *Chem. Soc. Jpn.* 2021, **94**, 106.
45. L. Chen, S. Yang, Q. Zhang, J. Zhu, and P. Zhao, *Sep. Purif. Technol.* 2021, **265**, 118444.
46. F. Ghanbari and M. Moradi, *Chem. Eng. J.* 2017, **310**, 41.
47. L. Zou, X. Xiao, C. Chu, and B. Chen, *Sci. Total Environ.* 2021, **775**, 143398.
48. S. He, R. Yin, Y. Chen, T. Lai, W. Guo, L. Zeng, and M. Zhu, *Chem. Eng. J.* 2021, **423**, 130172.
49. Q. Yi, J. Tan, W. Liu, H. Lu, M. Xing, and J. Zhang, *Chem. Eng. J.* 2020, **400**, 125965.
50. S. Dong, L. Xia, X. Chen, L. Cui, W. Zhu, Z. Lu, J. Sun, and M. Fan, *Compos. Part B Eng.* 2021, **215**, 108765.
51. A. Rezaei, M. R. Rezaei, and M. H. Sayadi, *J. Taiwan Inst. Chem. Eng.* 2021, **121**, 154–167.
52. R. Liu, Z. Chen, Y. Yao, Y. Li, W. A. Cheema, D. Wang, and S. Zhu, *RSC Adv.* 2020, **10**, 29408.
53. J. Liu, X. Wei, W. Sun, X. Guan, X. Zheng, and J. Li, *Environ. Res.* 2021, **197**, 111136.
54. R. -F. Guo, P. Liang, X. -Y. Li, and Z. -H. Liu, *Sep. Purif. Technol.* 2021, **264**, 118414.
55. Y. Yang, L. Xu, H. Shen, and J. Wang, *Sci. Total Environ.* 2021, **780**, 146576.
56. R. Zhang, W. Wan, D. Li, F. Dong, and Y. Zhou, 2017, *Chin. J. Catal.* **38**, 313.
57. P. Singh, P. Shandilya, P. Raizada, A. Sudhaik, A. Rahmani-Sani, and A. Hosseini-Bandegharaei, *Arab. J. Chem.* 2020, **13**, 3498.
58. J. Yin, D. Gao, X. Zhu, X. Liu, and H. Li, *Ceram. Int.* 2021, **47**, 19556–19566.
59. Y. Song, Y. Peng, N. V. Long, Z. Huang, and Y. Yang, *Appl. Surf. Sci.* 2021, **542**, 148584.
60. T. Xiong, Y. Ye, B. Luo, L. Shen, D. Wang, M. Fan, and Z. Gong, *Ceram. Int.* 2021, **47**, 14290.
61. S. Dong, L. Cui, C. Liu, F. Zhang, K. Li, L. Xia, X. Su, J. Feng, Y. Zhu, and J. Sun, *J. Taiwan Inst. Chem. Eng.* 2019, **97**, 288.
62. H. He, L. Huang, Z. Zhong, and S. Tan, *Appl. Surf. Sci.* 2018, **441**, 285.

63. J. -Y. Mei, P. Qi, X. -N. Wei, X. -C. Zheng, Q. Wang, and X. -X. Guan, *Mater. Res. Bull.* 2019, **109**, 141.
64. L. Lin, Q. Xie, M. Zhang, C. Liu, Y. Zhang, G. Wang, P. Zou, J. Zeng, H. Chen, and M. Zhao, *Colloids Surfaces A Physicochem. Eng. Asp.* 2020, **601**, 124978.
65. C. Kim, K. M. Cho, K. Park, K. H. Kim, I. Gereige, and H. Jung, *Chempluschem* 2020, **85**, 169.
66. C. Ding, Z. Li, W. Tan, H. Li, J. Ma, Z. Chen, Y. Tao, Y. Qin, and Y. Kong, *Synth. Met.* 2018, **246**, 137.
67. Z. Bano, R. M. Y. Saeed, S. Zhu, M. Xia, S. Mao, W. Lei, and F. Wang, *Chemosphere* 2020, **246**, 125846.
68. Y. Jiang, S. Chowdhury, and R. Balasubramanian, *J. Colloid Interface Sci.* 2019, **534**, 574.
69. Q. Liang, S. Ploychompoo, J. Chen, T. Zhou, and H. Luo, *Chem. Eng. J.* 2020, **384**, 123256.
70. W. Xiao, W. Zhou, Y. Zhang, L. Tian, H. Liu, and Y. Pu, *J. Nanomater.* 2016, **2016**, 1.
71. Y. Huang, C. Zhu, H. Pan, D. Xu, T. Lu, L. Mao, X. Meng, Z. Chen, D. Zhang, and S. Zhu, *RSC Adv.* 2017, **7**, 36000.
72. S. Jin, Y. Yang, J. Zhang, and H. Zheng, *Mater. Chem. Phys.* 2021, **263**, 124339.
73. M. Zhang, Y. Chen, B. Chen, Y. Zhang, L. Lin, X. Han, P. Zou, G. Wang, J. Zeng, and M. Zhao, *New J. Chem.* 2019, **43**, 5088.
74. Y. Chen, Y. Liang, M. Zhao, Y. Wang, L. Zhang, Y. Jiang, G. Wang, P. Zou, J. Zeng, and Y. Zhang, *Ind. Eng. Chem. Res.* 2019, **58**, 3538.
75. X. Xin, S. Li, N. Zhang, Z. Tang, and Y. Xu, *Appl. Catal. B Environ.* 2019, **245**, 343.
76. A. L. T. Zheng, S. Sabidi, T. Ohno, T. Maeda, and Y. Andou, *Chemosphere* 2022, **286**, 131731.
77. X. Ma, Z. Wang, H. Yang, Y. Zhang, Z. Zhang, H. Lin, J. Long, X. Wang, and Q. Lin, *RSC Adv.* 2021, **11**, 20446.
78. M. Karbasi, F. Karimzadeh, K. Raeissi, S. Giannakis, and C. Pulgarin, *Chem. Eng. J.* 2020, **396**, 125189.
79. A. Masud, C. Zhou, and N. Aich, *Environ. Sci. Nano* 2021, **8**, 399.

# Chapter 4

# Green Nanomaterials Industrial Utilizations in Nanomedicine and Pharmaceuticals

**Praseetha P. Nair[a] and Prabir Patra[b]**
[a]*Department of Chemical Engineering, Government Engineering College, Thrissur, Kerala, India*
[b]*Department of Biomedical Engineering and Mechanical Engineering, University of Bridgeport, Bridgeport, Connecticut, USA*
praseetha@gectcr.ac.in, ppatra@bridgeport.edu

Nanomaterials (NMs) have exotic biological and physicochemical properties that make them ideal for numerous prospective applications in diverse fields. The contributions of nanotechnology in medicine are promising and its huge potentials remain unexplored. Green NMs (GNMs) extracted from biological sources like plant's extracts and microbes, develop effective treatment stratagems predominantly due to their biocompatibility, bioavailability, and less toxic nature. The antimicrobial properties of NMs can efficiently be exploited for biomedical applications. This chapter highlights the importance and scope of GNMs in nanomedicine and pharmaceuticals and the antimicrobial activity of NMs. Superior therapeutic systems with the application of GNMs in targeted and

*Green Nanoarchitectonics*
Edited by Kaushik Pal

ISBN 978-981-4968-50-8 (Hardcover), 978-1-003-31860-6 (eBook)
www.jennystanford.com

sophisticated drug delivery, theranostics, regenerative medicine, implantable devices, tissue reconstruction, etc., are analyzed against conventional therapeutic system. The chapter also gives remarkable significance on the discussion of the future scopes of nanomedicine as well as the challenges and feasibilities during the clinical translation of nanomedicine and pharmaceuticals that are key to the successful implementation of nanomedical drugs.

## 4.1 Introduction and Background

The supremacy of nanotechnology is witnessing explosive advancements in diverse interdisciplinary arenas. Nanotechnology engraved its footprints in all walks of life productively owing to the exceptional properties of nanomaterials (NMs). Nanomedicine, the amalgamation of nanotechnology with medicine, paves the way for revolutionary advancements in conventional medical systems. The speculations about the enormous prospects of nanomedicine seems to be fanciful, but nanomedicine could make its mark in history by its numerous apparent realities.

### 4.1.1 Historical Background

Our nature is a plentiful resource of many medicinal herbs that have been used from time immemorial. Nature influences drug discovery and its application in therapy. Most of the drugs and pharmaceuticals that we are using today are derivatives of natural products mainly from flora and fauna. The use of natural drugs is less expensive as no downstream processing is required for its purification. However, the efficient application of extracts from nature is limited in most circumstances especially for the treatment of complex diseases due to several downsides of the purest form of drugs. The bioavailability of such drugs to its fullest extent is restricted as they are sparingly soluble, less diffusive, and less absorptive into the body parts. Bioavailability of a drug is the accessibility of its required constituents to the region of action of the body. Moreover, there are several other drawbacks like the requirement of heavy and frequent doses, less shelf life, lack of stability, etc., that deteriorate their performance during administration. The manufacture of manmade drugs could improve their properties thereby enhance the performance remarkably that leads to the traditional medicinal system. The properties of materials and methodologies prevailing in our conventional medical systems become advanced with the incorporation of green nanotechnology. Despite the vastness of the topic, nanomedicine, the current chapter highlights the significant biomedical and pharmaceutical applications of green NMs (GNMs), which revolutionize the medical field and pave the way for future therapeutic approaches on chronic ailments. This chapter also provides a focused overview of the characteristic features, antimicrobial action, and mechanism of GNMs and the advantages and challenges of nanotechnology in medicine.

### 4.1.2 Nanotechnology in Medicine

Nanomedicine is the tool of nanotechnology, which can create wonders in the field of modern medicine. NMs are very small particles similar in size as that of biological entities which make them able to interact with cells, proteins, DNA, enzymes, blood vessels, cell receptors, etc. The principle of nanomedicine is the engineered use of NMs in various aspects of medicine by understanding the biological processes in the system. Most of the complex diseases are diagnosed in the fatal stage of a patient that leads to failure in therapy. Early stage detection of diseases, targeted drug delivery, and adequate therapy are possible through the avenues of nanomedicine [34]. Diagnosis at the primary stage can initiate proper treatment strategies. Nano-engineered schemes can be generated, that are able to identify diseased cells and destroy them without affecting normal cells, compared to conventional therapies. Targeted drug delivery can use sophisticated targeting with NMs to penetrate into cells and kill or repair only the sick ones without affecting the healthy cells. Looped control systems can be implemented for cell targeting and drug delivery [5]. Another evolving area is regenerative medicine where techniques can be developed for the replacement or maintenance of tissues, cells, or organs. Cell proliferation can also be sustained through this branch of medicine through engineered NMs. The applications of nanomedicine in various aspects are listed in Table 4.1.

**Table 4.1** Applications of nanomedicine

| Area | Description |
|---|---|
| Early stage diagnosis | Leads to proper treatment strategies for complex diseases like cancer |
| Targeted drug delivery | Can be used to target unhealthy cells leaving behind healthy cells |
| Regenerative medicine | Can be used for the replacement or maintenance of diseased cells, tissues, or organs |
| Nanorobots | Can be used to assess diseases, deliver drugs, repair or replace cells or tissues, correct genetic deficiency, DNA sequencing, etc. |
| Slow and controlled release of drugs using nanomachines | Can be used as carrier to avoid heavy and frequent doses of medicine. |
| Development of new engineered novel nanostructures | Can be used in all fields from diagnosis to therapy to enhance conventional medical devices. |
| Development of new techniques | Can be used for delivery of drugs, tissue engineering, treatment of chronic diseases, etc. |
| Development of nanodevices | Can be used for implants and tissue construction |
| Development of contact lenses | Can be used to develop contact lenses for in situ drug distribution |
| Enhancement of bioavailability of drugs | Solubility of the drugs can be improved by administering them as nanopowders. |
| Development of nanosensors | Can be used for with-patient testing and analysis that gives fast real time analysis. |
| Personalized medicine | Patient specific advanced treatment methods can be formulated with the help of genomics and proteomics. |
| Vetinary science | Early diagnostics of diseases and smart delivery of drugs using nanosensors combined with monitoring and control, can be implemented. |
| Nanoformulations | Tools like nanocoatings, doping, emulsions, etc., can be adopted to implement nanomedicine effectively. |

## 4.2 Green NMs

### 4.2.1 An Insight into GNMs

GNMs are materials that are synthesized through environmental friendly and sustainable techniques with the principles of green chemistry and green nanotechnology with nontoxic materials. Materials synthesis processes should be designed such that the employment of green manufacturing and eco-friendly techniques are ensured. Existing physical or chemical methods of manufacturing of NMs are quite unsustainable. The requirement of extreme controlled atmosphere and rigorous operating conditions for production of NMs makes it high-energy and capital-intensive process. It also lead to the production of hazardous residues and demands further downstream processes for refining, ultimately resulting in poor yield.

**Table 4.2** Principles of green chemistry

| **Principles of green chemistry** | Waste reduction | Process safety | Materials efficiency | Energy efficiency | Customized nanomaterials |
|---|---|---|---|---|---|
| Prevent waste | ✓ | | ✓ | | |
| Maximize atom economy | | | ✓ | | |
| Less hazardous chemical synthesis | | ✓ | | | |
| Safer chemicals and products | | ✓ | | | ✓ |
| Safer solvent/ reaction media | ✓ | ✓ | | | |
| Increase energy efficiency | | | | ✓ | |
| Use renewable feed stocks | ✓ | | ✓ | | |
| Reduce derivatives | ✓ | | ✓ | ✓ | |
| Use catalysts | ✓ | ✓ | ✓ | ✓ | ✓ |
| Design for degradation | ✓ | | ✓ | | ✓ |
| Real-time monitoring and process control | ✓ | ✓ | ✓ | ✓ | ✓ |
| Inherently safer chemistry | | ✓ | | | |
| | ***Design strategies for green synthesis of nanomaterials*** | | | | |

*Source*: Ref. 30 (Copyright @ Elsevier, 2018).

Greener routes are employed due to the ability of NMs to form self-assembled monolayers and also due to their enhanced interaction with biomolecules. When the principles of green chemistry are applied for synthesis, process conditions will be less stringent, waste generation is less, purification steps and energy requirements are reduced, yield will be maximized, production rate is enhanced, and long-term sustainability is ensured. The 12 principles of green chemistry, that can be applied to design sustainable techniques for NMs synthesis, are summarized in Table 4.2 [30]. Since the mode of operation is flexible for green synthesis, control can be applied for the production of monodisperse NMs.

### 4.2.2 Advantages of Green Synthesis

Green synthesis routes are followed for material production where natural and biological materials like plants, animals, and microbes are used. These are less expensive and environmentally sustainable methods that reduce the generation of waste, produce less toxic materials with high stability, controlled size, shape, and bioactivity [4]. Biotemplate-assisted synthesis is a well-accepted, rapid, cost-effective, and eco-friendly technique for the production of GNMs. Large-scale production limits the biological synthesis in certain circumstances. Thus, an alternative green synthesis route can be adopted which uses biological methods combined with chemical principles and green chemistry. Metallic nanoparticles like silver, gold, iron, etc., are very demanding for biomedical applications due to their medicinal properties besides small size and high specific surface area. New class of cellulose-based NMs are synthesized from biomass and byproducts from pulp and paper industry. Its large-scale production is possible through less expensive processes using plant-based cellulose sources. Cellulose NMs for medical applications are synthesized via microbial route because of its biocompatibility, even though rate of production is low. Besides microbial precursors, algae, plants, their extracts, aquatic and terrestrial weeds, etc., have been used for biosynthesis of NMs. Polyphenols in plant extracts help in the degradation of organic precursors to form NMs. Cost-effective and eco-friendly synthesis of perovskite Zn–Sn–O NPs were reported by Mayedwa et al. [23] using *Aspalathus Linearis* plant extracts. Leaves and fruits of green plants are used for the synthesis of iron NPs [6].

Microbes are used for the intracellular and extracellular synthesis of AgNPs [38]. Strains of *Microbacterium marinilacus, Aspergillus niger, Pleurotus* sp., etc., are used for the synthesis of iron NPs [6].

### 4.2.3 Properties of NMs Applicable to Medical Field

Bioavailability and biocompatibility of materials are the essential requirements for biomedical applications especially in tissue engineering. Carbon-based NMs like fullerenes, graphene, carbon nanotubes (CNTs) have outstanding mechanical strength, good thermal and electrical conductivity. Their properties can be enhanced when used in combination with conventional materials. The properties of the biomaterials impact the end uses of carbon NMs (CNMs). Graphene, its derivatives, and functionalized groups have great biocompatibility. Besides that, they possess excellent optical properties, electrical conductivity, electronic properties, and electrocatalytic activity [19]. Studies conducted by Aggarwal and coworkers showed that reduced graphene oxide (rGO) has effective interaction with living cells. Due to the presence of sp2 hybridized carbon atoms and the very high specific surface area, graphene can adsorb or conjugate with molecules that makes them suitable as drug carriers [43]. Graphene oxide is also used as drug carriers for cancer due to its high drug loading capacity. CNMs possess vital characteristics of drug carriers such as controlled delivery, loading capacity, multiple drug delivery, etc. Multidrug resistance is a serious issue shown by certain microbes against any one antibiotic during multiple drug delivery to certain complex disease like tumor. Studies show that CNMs can be used to overcome the multidrug resistance, by carrying therapeutic drugs along with a drug resistance inhibitor [12] which can target the affected cells. CNTs consist of carbon atoms arranged in hexagonal lattice with high aspect ratio and hence, as graphene, they are suitable for biomedical applications like drug delivery, bioimaging, as nano-biosensors, etc. Multi-walled CNTs (MWCNTs) are made up of concentric cylinders of very small diameters that can capture drug molecules perfectly during the delivery of drugs, while there are many studies that reported the use of single-walled CNTs (SWCNTs) also as drug carriers. CNTs, especially SWCNTs, have very good fluorescent properties that makes them suitable for bioimaging.

### 4.2.4 Various Methods of Modification of NMs for Nanomedicine

The effectiveness of the biomedical usage of NMs accounts for its ability to interact appropriately with biomolecules than particle targeting. Agglomeration, hydrophobicity or hydrophilicity, solubility, etc., can hinder the direct contact between interfaces and limits the penetration capability of nanomedicine. Surface modification of NMs is necessary to fully harness their properties for suitable applications. Coating of core NMs with suitable shell materials is a common surface modification method. Noncovalent forces like electrostatic attraction and hydrogen bonding helps in the physical adsorption of coatings, whereas covalent bond helps in chemical adsorption. Coatings enhance the stability and monodispersity of NMs and prevents agglomeration. The properties of magnetic nanoparticles can be suitably tuned by coating with organic or inorganic polymers like polyvinyl alcohol, polyethylene glycol, chitosan, dextran, etc. Polyethylene glycol and polyethylene imine are used as coatings for CNTs [40]. Metal nanoparticles like gold and silver also act as coating materials. The properties of these surface-coated NMs can further be improved by functionalization with antibodies, peptides, polysaccharides, etc. Functionalized NMs exhibit improved interactions with cells, ions, drugs, etc. Further derivatization is also easy with functionalization. Functionalized molecules can be coupled with other molecules like polysaccharides, during surface treatment. Another surface treatment method is doping with atoms of nitrogen and sulfur. Doping of quantum dots with nitrogen or sulfur enhances their stability and sensing capacities. Preparation of composite materials from various NMs also improves their selectivity and specificity [8] for bioimaging and drug delivery.

## 4.3 Antimicrobial Activities of GNMs

NMs exhibit antimicrobial reaction to many pathogens. Their antimicrobial activity can be enhanced by surface modification techniques. The factors like specific surface area, surface to volume ratio, size, shape, quantity, nature of nanoparticle and its stability, nature of pathogen, mode of interaction, contact time, etc. [39], influence the antimicrobial activity. Carbon-based NMs like fullerenes, graphite, CNTs, graphene, and their derivatives like graphene oxide (GO) and graphene quantum dots (GQDs) are profound in their antimicrobial activity due to their exotic properties. Other green inorganic NMs include silver and gold particles and currently relevant NMs based on ZnO, $TiO_2$, and cellulose.

## 4.3.1 Effect of GNMs on Microbes and Their Biomedical Applications

### 4.3.1.1 Carbon nanomaterials

There are different approaches for the green synthesis of CNMs using biological precursors like rice husk, biomass, etc. [2, 26, 28]. One of the most commonly used CNMs is CNTs that fall under two categories, namely, SWCNTs and MWCNTs. Both differ in their antimicrobial effect toward various microbes. The toxic effect of CNTs depends on the length of their chain [27]. CNTs in stipulated concentrations [3] can be used to reduce the bacterial content in soil. MWCNTs have less effect on fungi [36]. Surface treatments like functionalization, acid treatments, etc., enhance their toxic property [26, 35]. SWCNTs and MWCNTs show toxic effect in aquatic systems and hence this property can be better utilized in water treatment systems. Fullerenes are highly toxic to microorganisms as well as plants [21]. Their toxicity depends on the type of microbe and the concentration of fullerene while they are less toxic toward aquatic and anaerobic microorganisms [33]. Graphene shows higher biotoxicity compared to other CNMs even at low concentrations. There is a positive effect on the presence of graphene in soil at low concentrations that favors the growth of valuable bacteria in soil, namely, nitrogen fixing bacteria [11]. Having high toxicity, graphene derivatives and composites attack microbial cells leading to cell death.

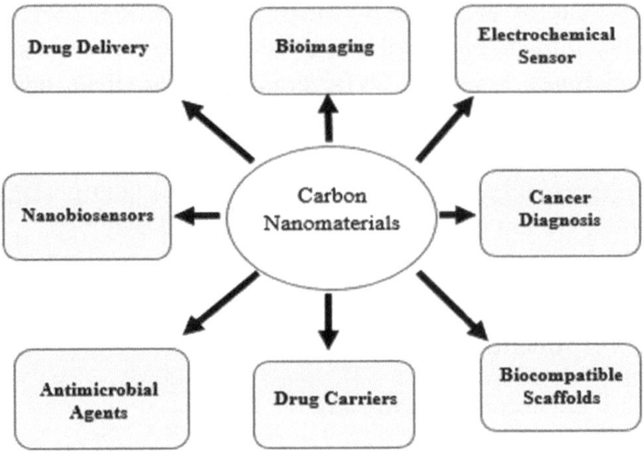


**Figure 4.1** Biomedical applications of CNMs.

The most common areas where CNMs find their biomedical application are in bioimaging and nano-biosensors. Graphene is the most versatile among them, which is used in biomedical imaging, sensing, drug delivery, cancer treatment, biocompatible scaffold, and microbial inhibitor. GO and GQDs are used as sensors [32]. GQDs are used in tissue imaging. Graphene is ideal for cancer diagnosis and also as nanocarrier for drugs into targeted sites [43]. Due to its high and controlled drug loading capacity, GO is also used as drug carrier for cancer. Both SWCNTs and MWCNTs are used as effective drug carriers. Graphene and its derivatives are used for bioimaging as they can carry fluorescent species. Composites based on graphene can be used as biosensors for the detection of hemoglobin, glucose, and its compounds in our body [3]. Li et al. [22] reported the production of electrochemical sensor based on GQD and gold nanocomposite. The biomedical applications of carbon-based NMs are summarized in Fig. 4.1.

#### 4.3.1.2 Nanosilver

The antimicrobial properties of silver has been known from ancient times. Silver has been used during world wars in antiseptic ointments to heal injuries. Antimicrobial characteristics of silver nanoparticles (SNPs) make it very much appropriate for medicinal applications and also in waste water disinfection. It has demand in cosmetic industry due to the ability of scarless healing of wounds.

The antimicrobial action of SNPs is influenced by its size, shape, specific surface area, concentration, and zeta potential. Superior toxicity is exhibited by very small particles as cellular penetration is easy as size is reduced. There are many contradictory statements regarding the effect of shape of SNPs on toxicity. El-Zahry et al. [10] showed that shape affects toxicity, while Actis et al. [1] reported that shape has no influence. The strength of toxicity of SNPs toward various bacterial strains were analyzed by Kim et al. [18] and found inhibition to *Escherichia coli* at mild concentration of NPs, but toxic effects are less to *Staphylococcus aureus*. A combination of SNPs and titanium dioxide NPs are toxic toward fungi than bacteria [14]. Methanogenic bacteria is affected by certain concentration of SNPs. The nitrifying bacteria present in waste water and soil are affected by SNPs and polyvinyl pyrrolidone-coated SNPs [7]. As the dosage of SNPs increases, sustainability of cell decreases. But this cannot be generalized and is specific to microbial strain. Electrostatic adhesion contributes to microbe–nanoparticle interaction and hence zeta potential plays a crucial role in the antimicrobial activity of SNPs [31]. Positively charged SNPs get adhered to negatively charged cells of microbes and initiates toxicity.

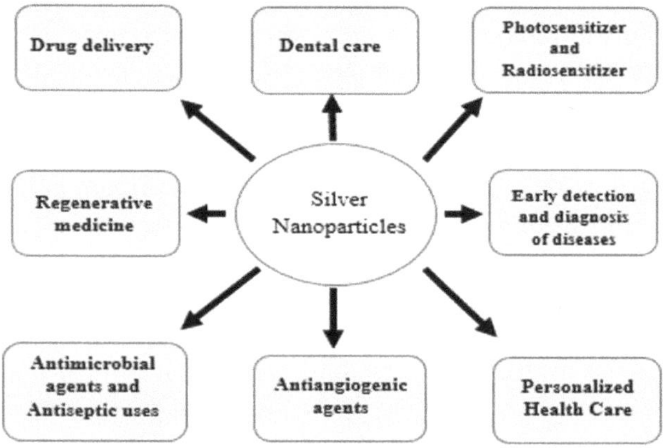


**Figure 4.2** Biomedical applications of SNPs.

The biomedical application of SNPs ranges from diagnosis to therapeutics. Due to their high biocompatibility, they are effective drug carriers especially for anticancer and antitumor drugs. SNPs incorporated materials and medical devices are used in clinics due to their antimicrobial characteristics. Toxicity to microbes makes them suitable to apply for healing of wounds. Bioinspired bone scaffolds made from SNPs can be used for dental care applications such as antimicrobial mouth washes, dental implants, intraoral prosthesis, fillings for root canal treatment, etc. Since they are capable of inducing photosensitivity, they are being used as photosensitizer in photo chemotherapy. Similarly, due to its ability to sensitize tumorous and cancerous cells toward radiation, they can be used as radiosensitizer in radiation therapy. Moreover, SNPs can used as antiangiogenic agents and anticancer agents especially for lung cancer, breast cancer, skin cancer, liver cancer, and leukemia [41]. A brief summary of the biomedical application of SNPs is given in Fig. 4.2.

#### 4.3.1.3 Currently relevant NMs for biomedical applications

Gold NPs and their functionalized forms have extensive biomedical applications due to its inert nature and biocompatibility. They are used in drug delivery, bioimaging, radio sensitization, and in gene therapy. The surface plasmon resonance exhibited by gold NPs is exploited for treating cancer by photo thermal therapy. Localized heat developed by the absorbed light is used to eradicate cancer cells without subjecting the cells to any radiation. ZnO NPs have limitations in biomedical usage as they produce toxic byproducts. However, green synthesized ZnO NPs exhibit cytocompatibility and biomedical activity. It is excellent in killing cancer tissues by generating reactive oxygen species (ROS) during contact [7]. ZnO doped with QDs is used as drug carriers in cancer treatment. ZnO NPs also exhibit wound-healing properties, osteogenesis, angiogenesis and antineoplastic effects [17]. This makes them suitable for regenerative medicinal and tissue engineering applications. As ZnO, nanostructures based on $TiO_2$ are effective in tissue regeneration and diagnosis of diseases like diabetes, orthopedic and cardiovascular disorders. They are also used in medical implants like dental, vascular stents, orthopedics and also in drug delivery system [42]. Due to its photocatalytic activity, $TiO_2$ is used for the photocatalytic deactivation of microorganisms. It can be used as toxic agent to many bacteria like gam-positive and gram-negative as well as viruses like hepatitis B, polio, etc. Cellulose based products and their derivatives have abundant use in diverse fields. The medicinal applications of cellulosic NMs include medical implants, drug delivery, maintenance and replacement of tissues as they are highly biodegradable, biocompatible and less toxic to cells [25]. Cellulose based hydrogels in tissue maintenance provides an active and fast repair, healing and regeneration of tissues. They are also used for tissue replacement in cartilages, heart valves, etc. Cellulose based hydrogel bandages are compatible to skin and possess bacterial inhibition. They are used in medical fields that promise rapid healing of wounds. Another class of hybrid NMs that are formed by the combination of both organic and inorganic components like silica based materials, finds application in imaging and therapy [29]. NMs having photophysical and luminescence properties are used for biomedical imaging, theranostics and biosensing [9].

### 4.3.2 Mechanism of Antimicrobial Activity of GNMs

Appropriate understanding of the mechanism of the antimicrobial activity of NPs is essential for exploring their applications in various fields. Many researches were conducted for elucidating the mechanism of the toxic effect of NPs on microorganisms. Most of the hypothesis are specific to certain microbes, while some are common to all. As NPs are extremely small in size it can easily diffuse or penetrate through the cell membrane leading to the rupture of cell wall or membrane. This can destroy proteins, DNA, nucleic acid, etc. These will damage all functions of cell and will interfere energy transduction process, eventually lead to the death of microbial cell. Oxidative stress caused by the excessive production of ROS in NMs exerts toxic activity to them against microbes. NMs interact with cellular constituents like proteins and deactivate the enzymes that produce ROS or metal ions. Oxidative stress causes an imbalance between ROS and antioxidant mechanism. As a response to overcome this stress, cells exhibit self-defense mechanisms. When oxidative stress overpowers these mechanisms, damage happens to the constituents due to ROS. These can totally disorder the cell structure, inhibit DNA replication, and produce many harmful radicals that are toxic to microorganisms [15]. SNPs having size below 10 nm is superior in their toxicity due to their easy penetration into the cellular structure, while toxic effect of other sizes are due to the release of silver ions [33]. NMs will be adhered to microbial cells and causes structural changes of cell membrane, intracellular penetration and damage, and increase its permeability, leading to the leakage of cellular constituents and eventually causes cell death [33]. The various possible antimicrobial mechanisms of NMs are summarized in Fig. 4.3 [15].

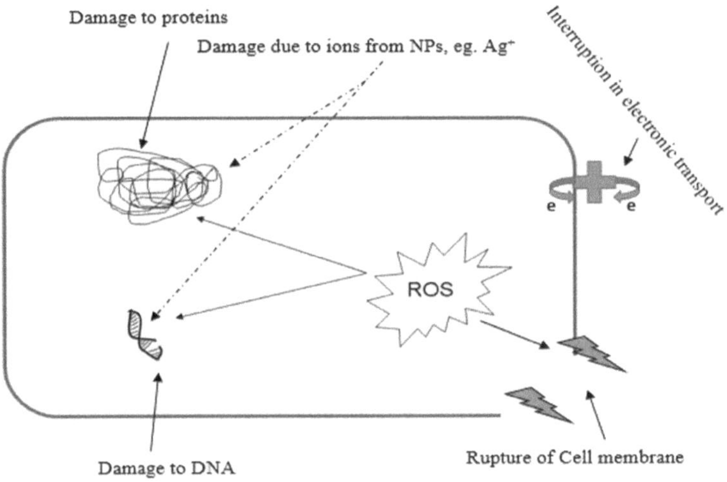


**Figure 4.3** Antimicrobial mechanisms of nanoparticles. *Source*: Modified image from Ref. 15 (Copyright @ Elsevier).

## 4.4 GNMs Approaches to Nanomedicine and Pharmaceuticals

The treatment of deadly diseases like cancers and tumors has inconsistent and unpredictable outcomes in conventional medical therapy. The approaches for treating these dreadful disorders in the conventional treatment system are radiation, chemotherapy, or removal of the affected cells or part. Nanomedicine follows a smart therapeutic approach of repairing or replacing the affected cells or organs based on the diagnosis. Many unsolved clinical and pharmacological problems such as chemotherapy or radiotherapy resistant tumors have solutions in nanomedicine. Major disadvantages of treatments with current system for brain and spinal tumors are side effects like drug accumulation due to heavy and unspecific dosage of drugs and neurotoxic problems [14]. The usage of NMs also makes the blood–brain barrier permeable to many drugs. The biomedical and pharmaceutical applications of NMs are concentrated in three main areas: theranostics which is a combined approach of diagnosis of diseases and therapy, targeted and controlled drug delivery, regenerative medicine and tissue engineering. The green synthesis makes NMs biocompatible and biodegradable, thus suitable for safer use as carriers of medicines.

### 4.4.1 NMs in Theranostics

Medical field is facing challenges of early detection of many ailments that can make therapies easier. NPs as biomarkers are used as indicators to identify any normal or abnormal biological processes that help in the detection of a state of disorder in the body and are used for the early stage diagnosis and prognostics of chronic diseases like malignant tumors and cancers. Metallic NPs like gold and silver, carbon-based NMs, quantum dots, etc., can be used for the detection of proteins and DNA for the diagnosis and therapy of diseases. Cancers are really disastrous which are usually identified at the fatal stage. This situation is alleviated with the prospective use of nanomedicine. Primary stage diagnosis with nano aids leads to a successful treatment of cancer by killing the sick cells and maintaining the healthy cells. Further spread of cancerous tissues can also be arrested with effective therapy in nanomedicine. Magnetic iron oxide NPs developed for medical and neuroimaging can give detailed cellular and molecular imaging for the detection of cancer. The magnetic iron NPs can cause oxidative stress that produces ROS and free radicals, induce antimicrobial action leading to the cell death [14]. Manganese oxide NPs are used in bioimaging to produce bright images that aid in targeting. Metal nanoshells and quantum dots produce high resolution images of malignancies. Polystyrene nanostructured surfaces were fabricated for the detection of protein that helps in clinical diagnosis by Kunduru et al. [20].

### 4.4.2 Drug Delivery System

Site specific drug delivery is the effective way of applying antitumor drugs with desired pharmacological effects. Surface-modified NPs conjugated with antibodies are used as pharmaceutical drugs due to their specificity, nontoxicity, biodegradability, and stability. Dendrimers are used as carriers of drugs and are conjugated to form ligands for permeating through blood–brain barrier [14]. They can also be suitably used for tumor targeting without affecting healthy cells. Biocompatible and biodegradable polymeric NMs are suitable for drug delivery. Slow release of drugs ensures its prolonged bioavailability with controlled doses [37]. Cross-linked polymeric networks of hydrogels can be used as a promising drug delivery vehicle due to its improved solubility. Nano compact dendrimers are suitable for carrying drugs during administration due to their ability to perform sustained and targeted drug delivery. Similarly, micelles can be used as drug carriers by incorporating with hydrophobic drugs. Drugs-encapsulated polymeric NPs can be utilized for drug delivery as they can form binding with macromolecules. Paclitaxel drug delivery for effective treatment of tumor was studied using nanoformulated water-soluble drug with minimized side effects [13]. Inorganic NMs of gold, silver, zinc, copper, etc., are used in various pharmaceutical studies. Hydrophilic and hydrophobic drugs encapsulated in liposomes are used for the controlled and targeted release of drugs due to their ability to inhibit the prompt decomposition of drugs. Liposomes and their functionalized derivatives like chitosan-coated liposome show great potential in entrapping many drugs, especially insulin. They facilitate effective diabetic treatment with less insulin leakage and site-specific drug administration. Neurological disorders like Alzheimer and Parkinson can be therapeutically treated with nanodrug delivery system. The inefficiency of drug molecules to pass through the blood–brain barrier seriously limits the therapy of neurodegenerative disorders. This can rightly be surmounted with suitable nanodrugs or carriers. Moreover, nanomedicine is capable of sustained release of medicines that makes the drug bioavailable for long time and it reduces the requirement of frequent and heavy dosages and minimizes side effects.

### 4.4.3 Regenerative Medicine and Tissue Engineering

Regenerative medicine is a therapy of restoring the damaged cells, organs, and tissues by repair or replacement so as to reinstate the normal structure and function. This can be done either through self-repair by the body or by organelles implantation after the in vitro growth of organs and tissues, if it cannot be self-regenerated naturally or by any stimulation mechanism. The method of regeneration through artificial organ or tissue transplantation is quite expensive as well as challenging for the availability of suitable donors. Another approach is stem cell therapy where magnetic NMs are used to isolate, sort, and trace stem cells. Gold nanorods and quantum dots are used for imaging and tracing of stem cells. CNTs are used to administer drugs to stem cells and for stem cell proliferation and differentiation. Another strategy is the development and implantation of biocompatible scaffolds in locations where new tissue is to be regenerated. Scaffolds aid in cell adhesion, proliferation, and differentiation and they are extensively used in bone regeneration [37]. Chitosan and cellulose bio-nanocomposites form attractive bio-scaffolds for tissue engineering owing to their excellent strength, ease of synthesis, and biocompatibility. Carbon-based NMs, polyaniline, gold NPs, etc., are used for developing biomimetic scaffolds for tissue engineering applications. Many cell-based medical devices and biopharmaceuticals for regenerative therapy are commercially available which are approved by Food and Drug Administration.

## 4.5 Challenges in Nanomedicine

Nanomedicine significantly influenced medical field by enhancing the treatment methods and therapeutic index of drugs and its delivery. However, the clinical transformation of nanodrugs poses severe challenge to the administration of nanomedicine. Some practical approaches should be developed to comprehend the in vitro results obtained to prognosticate and treat in vivo effects. The main factors that are to be answered during clinical translation are the generalization efficacy of the research outcomes in providing ideal biocompatibility and biodegradability to in vivo applications. Assessment or minimization of vulnerabilities in the development of nanomedicine and safety aspects during its administration are the key factors to be evaluated. Other crucial issues are the commercialization, large-scale production, and practical feasibility of administration of nanomedicine. Overall cost effectiveness and bioavailability should be ensured during clinical development [16]. Accumulation of nanodrugs at noncancer cells within the tumor sites should be properly addressed. Suitable techniques should be developed for nanoformulations for improving the efficiency of drug carriers. Further, voluminous researches on nanotoxicological studies to interpret the toxic paradigm of the chemotherapeutic effects of nanomedicine to human beings are to be conducted [24]. Issues on the impact of nanodrugs towards the immune system of human beings are also not properly answered. It is essential do studies on the evaluation of kinetic characteristics of nanomedicine that influences their pharmacokinetic properties (selective drug delivery, enhanced solubility, controlled release, improved permeability, biodegradability, increased exposure, and retention time) for better development of safer and effective nanomedicines. Government funding agencies and academic professional societies should promote extensive collaborative research between experts in pharmacology, nanomedicine, and toxicology for the implementation of nanomedicine by addressing all the encounters and challenges particularly in clinical translation.

## 4.6 Conclusion and Future Perspectives

Undoubtedly, the emergence of nanomedicine has extraordinarily influenced conventional medical approaches and made remarkable changes in the existing therapies. Many novel strategies like nanodrug carriers for targeted drug delivery, nanoformulations to enhance the properties of nanocarriers, regenerative medicine, stem cell therapy, etc., are discussed in detail in this chapter. There are unlimited and unexplored prospects of nanomedicine in cancer and tumor therapy. One of the most significant healthcare strategy is precision medicine or personalized therapy that works on patient-specific treatment or treating a group of people having common characteristics, with the help of genetic engineering in nanomedicine. This is relatively a costlier approach for pharmaceutical industry to accept it, in addition to various risks and challenges for developing personalized medicine. However, the scientific community is so hopeful that this concept will become the medicine of the future. Long-term researches have to be conducted along with pharmaceutical industries to discover the potentials of personalized medicine. The scope of bloodless and needleless surgery with the help of nanomedicine is also not out of question. Miniature surgeries are promising with nanorobots or nanobots that can go inside the body to do the repair of the damaged cells and tissues. The hereditary passage of diseases can be eradicated with nanobots by replacing the DNA molecules. Miniature electronic devices can be developed in the form of tablets with a video camera, when inserted inside body can take visuals and track drugs by sensing and imaging. The principles of nanomedicine can be used to develop nanoparticles that signal the presence of cancer cells in body at an early stage. New technologies for the development of sensors for remote monitoring of patients, healing systems for internal hemorrhages, etc., are under research. Indispensable steps should be taken to improve the clinical translation and significance. Despite the various challenges, the future generation can enjoy the potentials of nanomedicine in its fullest extent with committed research in this field.

### Ethical issues

This chapter does not contain any studies with human participants or animals performed by authors.

### Conflict of interest

The corresponding authors state that there is no conflict of interest.

### Acknowledgment

Copyright permission for reusing figures published by Elsevier is greatly acknowledged.

## References

1. Actis, L., Srinivasan, A., Lopez-Ribot, J. L., Ramasubramanian, A. K., and Ong, J. L. (2015). Effect of silver nanoparticle geometry on methicillin susceptible and resistant *Staphylococcus aureus*, and osteoblast viability. *Journal of Materials Science: Materials in Medicine*, **26**(7), 1–7.
2. Asnawi, M., Azhari, S., Hamidon, M. N., Ismail, I., and Helina, I. (2018). Synthesis of carbon nanomaterials from rice husk via microwave oven. *Journal of Nanomaterials*, **2018**, article ID 2898326, 5 p.
3. Barabadi, H., Ovais, M., Shinwari, Z. K., and Saravanan, M. (2017). Anti-cancer green bionanomaterials: present status and future prospects. *Green Chemistry Letters and Reviews*, **10**(4), 285–314.

4. Benelli, G. (2019). Green synthesis of nanomaterials. *Nanomaterials, 9,* 1275.

5. Bhowmik, D., Chiranjib, C. R., Tripathi, K. K., and Kumar, K. S. (2010). Nanomedicine: an overview. *International Journal of PharmTech Research,* **2**(4), 2143–2151.

6. Bolade, O. P., Williams, A. B., and Benson, N. U. (2020). Green synthesis of iron-based nanomaterials for environmental remediation: a review. *Environmental Nanotechnology, Monitoring and Management, 13,* 100279.

7. Cruz, D. M., Mostafavi, E., Vernet-Crua, A., Barabadi, H., Shah, V., Cholula-Díaz, J. L., Webster, T. J., et al. (2020). Green nanotechnology-based zinc oxide (ZnO) nanomaterials for biomedical applications: a review. *Journal of Physics: Materials,* **3**(3), 034005.

8. Devi, P., Saini, S., and Kim, K. H. (2019). The advanced role of carbon quantum dots in nanomedical applications. *Biosensors and Bioelectronics,* **141**, 111158.

9. Duan, C., Liang, L., Li, L., Zhang, R., and Xu, Z. P. (2018). Recent progress in upconversion luminescence nanomaterials for biomedical applications. *Journal of Materials Chemistry B,* **6**(2), 192–209.

10. El-Zahry, M. R., Mahmoud, A., Refaat, I. H., Mohamed, H. A., Bohlmann, H., and Lendl, B. (2015). Antibacterial effect of various shapes of silver nanoparticles monitored by SERS. *Talanta,* **138**, 183–189.

11. Freyria, F. S., Geobaldo, F., and Bonelli, B. (2018). Nanomaterials for the abatement of pharmaceuticals and personal care products from wastewater. *Applied Sciences,* **8**(2), 170.

12. Gong, M. Q., Wu, J. L., Chen, B., Zhuo, R. X., and Cheng, S. X. (2015). Self-assembled polymer/inorganic hybrid nanovesicles for multiple drug delivery to overcome drug resistance in cancer chemotherapy. *Langmuir,* **31**(18), 5115–5122.

13. Gu, W., Chen, J., Patra, P., Yang, X., Gu, Q., Wei, L., Acker, J.P., and Kong, B. (2017). Nanoformulated water-soluble paclitaxel to enhance drug efficacy and reduce hemolysis side effect. *Journal of biomaterials applications,* **32**(1), 66–73.

14. Hernández-Pedro, N. Y., Rangel-López, E., Magaña-Maldonado, R., de la Cruz, V. P., Santamaría del Angel, A., Pineda, B., and Sotelo, J. (2013). Application of nanoparticles on diagnosis and therapy in gliomas. *BioMed Research International,* 2013, article ID 351031, 20 p.

15. Hossain, F., Perales-Perez, O. J., Hwang, S., and Román, F. (2014). Antimicrobial nanomaterials as water disinfectant: applications, limitations and future perspectives. *Science of the Total Environment*, **466**, 1047–1059.

16. Hua, S., and Wu, S. Y. (2018). Editorial: advances and challenges in nanomedicine. *Frontiers in Pharmacology*, **9**, 1397.

17. Kalantari, K., Mostafavi, E., Afifi, A. M., Izadiyan, Z., Jahangirian, H., Rafiee-Moghaddam, R., and Webster, T. J. (2020). Wound dressing functionalized with silver nanoparticles: promises and pitfalls. *Nanoscale*, **12**(4), 2268–2291.

18. Kim, J. S., Kuk, E., Yu, K. N., Kim, J. H., Park, S. J., Lee, H. J., ... and Cho, M. H. (2007). Antimicrobial effects of silver nanoparticles. *Nanomedicine: Nanotechnology, Biology and Medicine*, **3**(1), 95–101.

19. Kuila, T., Bose, S., Khanra, P., Mishra, A. K., Kim, N. H., and Lee, J. H. (2011). Recent advances in graphene-based biosensors. *Biosensors and bioelectronics*, **26**(12), 4637–4648.

20. Kunduru, V., Bothara, M., Grosch, J., Sengupta, S., Patra, P. K., and Prasad, S. (2010). Nanostructured surfaces for enhanced protein detection toward clinical diagnostics. *Nanomedicine: Nanotechnology, Biology and Medicine*, **6**(5), 642–650.

21. Livingston, A., Trout, B. L., Horvath, I. T., Johnson, M. D., Vaccaro, L., Coronas, J., ... and Szekely, G. (2020). Challenges and directions for green chemical engineering: role of nanoscale materials. In *Sustainable Nanoscale Engineering* (pp. 1–18). Elsevier.

22. Maksimova, Y. G. (2019). Microorganisms and carbon nanotubes: interaction and applications. *Applied Biochemistry and Microbiology*, **55**(1), 1–12.

23. Mayedwa, N., Mongwaketsi, N., Khamlich, S., Kaviyarasu, K., Matinise, N., and Maaza, M. (2018). Green synthesis of zin tin oxide (ZnSnO3) nanoparticles using *Aspalathus Linearis* natural extracts: structural, morphological, optical and electrochemistry study. *Applied Surface Science*, **446**, 250–257.

24. Metselaar, J. M., and Lammers, T. (2020). Challenges in nanomedicine clinical translation. *Drug Delivery and Translational Research*, **10**(3), 721–725.

25. Moon, R. J., Schueneman, G. T., and Simonsen, J. (2016). Overview of cellulose nanomaterials, their capabilities and applications. *JOM*, **68**(9), 2383–2394.

26. Mugadza, K., Stark, A., Ndungu, P. G., and Nyamori, V. O. (2020). Synthesis of carbon nanomaterials from biomass utilizing ionic liquids for potential application in solar energy conversion and storage. *Materials*, **13**(18), 3945.
27. Nasimi, P., and Haidari, M. (2013). Medical use of nanoparticles: drug delivery and diagnosis diseases. *International Journal of Green Nanotechnology*, **1**(0), 1–5.
28. Osman, A. I., Farrell, C., Ala'a, H., Harrison, J., and Rooney, D. W. (2020). The production and application of carbon nanomaterials from high alkali silicate herbaceous biomass. *Scientific Reports*, **10**(1), 1–13.
29. Park, W., Shin, H., Choi, B., Rhim, W. K., Na, K., and Han, D. K. (2020). Advanced hybrid nanomaterials for biomedical applications. *Progress in Materials Science*, **114**, 100686.
30. Patwardhan, S. V., Manning, J. R., and Chiacchia, M. (2018). Bioinspired synthesis as a potential green method for the preparation of nanomaterials: opportunities and challenges. *Current Opinion in Green and Sustainable Chemistry*, **12**, 110–116.
31. Phanjom, P., and Ahmed, G. (2017). Effect of different physicochemical conditions on the synthesis of silver nanoparticles using fungal cell filtrate of *Aspergillus oryzae* (MTCC No. 1846) and their antibacterial effect. *Advances in Natural Sciences: Nanoscience and Nanotechnology*, **8**(4), 045016.
32. Qian, Z. S., Shan, X. Y., Chai, L. J., Chen, J. R., and Feng, H. (2015). A fluorescent nanosensor based on graphene quantum dots–aptamer probe and graphene oxide platform for detection of lead (II) ion. *Biosensors and Bioelectronics*, **68**, 225–231.
33. Roy, A., Bulut, O., Some, S., Mandal, A. K., and Yilmaz, M. D. (2019). Green synthesis of silver nanoparticles: biomolecule-nanoparticle organizations targeting antimicrobial activity. *RSC Advances*, **9**(5), 2673–2702.
34. Sandhiya, S., Dkhar, S. A., and Surendiran, A. (2009). Emerging trends of nanomedicine: an overview. *Fundamental and Clinical Pharmacology*, *23*(3), 263–269.
35. Shakeri, S., Ashrafizadeh, M., Zarrabi, A., Roghanian, R., Afshar, E. G., Pardakhty, A., ... and Thakur, V. K. (2020). Multifunctional polymeric nanoplatforms for brain diseases diagnosis, therapy and theranostics. *Biomedicines*, **8**(1), 13.
36. Sharma, D., and Hussain, C. M. (2020). Smart nanomaterials in pharmaceutical analysis. *Arabian Journal of Chemistry*, **13**(1), 3319–3343.

37. Singh, A. P., Biswas, A., Shukla, A., and Maiti, P. (2019). Targeted therapy in chronic diseases using nanomaterial-based drug delivery vehicles. *Signal Transduction and Targeted Therapy*, **4**(1), 1–21.

38. Subin P, Tapobrata P, Praseetha P N, and Thomas T (2014). Biosynthesis of silver nanoparticles. *Journal of Nanoscience and Nanotechnology*, 14, 2038–2049.

39. Uddin, I., Venkatachalam, S., Mukhopadhyay, A., and Amil Usmani, M. (2016). Nanomaterials in the pharmaceuticals: occurrence, behaviour and applications. *Current Pharmaceutical Design*, **22**(11), 1472–1484.

40. Wang, X., Liu, L. H., Ramstroem, O., and Yan, M. (2009). Engineering nanomaterial surfaces for biomedical applications. *Experimental Biology and Medicine*, **234**(10), 1128–1139.

41. Wei, L., Lu, J., Xu, H., Patel, A., Chen, Z. S., and Chen, G. (2015). Silver nanoparticles: synthesis, properties, and therapeutic applications. *Drug Discovery Today*, **20**(5), 595–601.

42. Wu, S., Weng, Z., Liu, X., Yeung, K. W. K., and Chu, P. K. (2014). Functionalized $TiO_2$ based nanomaterials for biomedical applications. *Advanced Functional Materials*, **24**(35), 5464–5481.

43. Zhang, L., Xia, J., Zhao, Q., Liu, L., and Zhang, Z. (2010). Functional graphene oxide as a nanocarrier for controlled loading and targeted delivery of mixed anticancer drugs. *Small*, **6**(4), 537–544.

# Chapter 5

# Green Nanomaterials in Photocatalysis Applications

**Meera A.P. and Reshma R. Pillai**
*Research and Post Graduate Department of Chemistry,*
*K. S. M. Devaswom Board College, Kollam, India*
apmeera@ksmdbc.ac.in

Nano photocatalysis is an emerging technology for environmental cleanup, which removes toxic substances through the acceleration of photoreactions by employing a catalyst. The balance of the ecosystem is being disturbed day-by-day by various anthropogenic activities and it is the need of the hour to have an eternal solution for the environmental remediation. Conventional methods involve chemical compounds which are environmentally toxic, whereas nanomaterials synthesized by the green route such as natural materials like plants, microorganisms, organic polymers, etc., are quite safe and ecofriendly. The main objective of this chapter is to present a systematic outline of the problem beginning with the mechanism of photocatalysis followed by the factors affecting the efficiency of a photocatalyst and the various green nanomaterials used for photocatalytic applications like metal nanoparticles, metal

*Green Nanoarchitectonics*
Edited by Kaushik Pal

ISBN 978-981-4968-50-8 (Hardcover), 978-1-003-31860-6 (eBook)
www.jennystanford.com

oxide nanoparticles, nanowires, nanorods, quantum dots, biopolymer coated nanoparticles, etc. are being used as photocatalysts.

## 5.1 Introduction

Nowadays both developed and developing countries have been facing serious problems of environmental pollution mainly due to anthropogenic activities. This badly affects the entire ecosystem including plants, animals, and humans. A huge increase in the number of industries and rapid urbanization led to environmental contamination to a greater extent [1]. There will be a great threat to the lives of future generations if the condition goes on like this. As a solution to this serious issue, several studies have been carried out in the area of environmental remediation, i.e., to reduce the pollutants from the environment such as air, water, and soil and to protect the environment from further deterioration [2].

Pollutants present in the environment are broadly divided into chemical pollutants and physical pollutants in which chemical pollutants are either organic or inorganic. Organic pollutants are those compounds that contain organic carbon which includes dyes, oils, insecticides, plastics, food additives, wastes from food and canning industries, paper and pulp mills. These wastes get into the environment either from the source of manufacture or from the use in different applications [3].

Among the common organic pollutants used, the most important one is the use of organic dye. About $7\times10^5$ tons of dyes are produced per year and a major percentage is released to the environment. This prevents the photosynthesis process of aquatic plants as it forms a layer on the water and also reduces the reoxygenation capacity of water [4]. Among the dyes used in the textile industry, the most used one is the Azo dye, which is carcinogenic in nature. When it is adsorbed on our skin, lungs, etc. causes serious health problems. Hence effective and rapid degradation of these dyes is one of the main challenges of researchers today. For the decontamination of water, several methods have been employed [5], some of which are listed in Fig. 5.1.

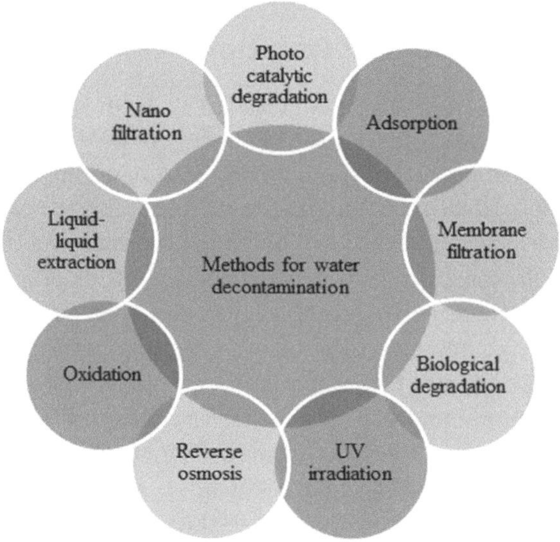


**Figure 5.1** Various methods for the decontamination of water.

Photocatalysis is a method of employing catalysts for increasing the rate of a reaction in the presence of light. Over other methods, photocatalysis possesses some advantages in which the reaction conditions are readily available and common in nature and the process gives harmless or pollution-free products and by-products. Thus photocatalysis has been intensively explored by scientists as a safe process to eliminate contaminants from the environment. Nowadays, nanomaterials are noted for their uniqueness in size, shape, crystal morphology as well as typical physical and chemical properties in comparison with their bulk counterparts [6].

Although various methods are adopted for the synthesis of nanomaterials, much attention has been paid to the method of green synthesis due to their non-toxic nature, low cost, high stability, and their environment-friendly behavior. Conventional methods involve toxic chemical compounds which cause environmental contamination. However, the green synthesis of nanomaterials using plants, microorganisms, and organic polymers is nontoxic and ecofriendly. This chapter gives a brief overview of the various aspects of photocatalysis and the scope of green nanomaterials in different photocatalytic applications.

## 5.2 Basic Principles and Mechanism of Photocatalytic Reactions

Heterogeneous photocatalysis uses solar light as the renewable source of energy and thus it is a cost-effective method. The result of a photocatalytic reaction is the mineralization of the pollutants [7]. In a photocatalytic reaction, the rate of reaction is found to depend mainly on the crystalline morphology and the energy of the incident photons. Photocatalysis involves redox reactions that depend on the electronic structure of the catalyst used. If the energy of the incident radiation is comparable to or greater than the band gap energy of the catalyst, excitation of electrons from the valence band (VB) to the conduction band (CB) occurs forms holes in the VB. Thus an electron-hole pair ($\bar{e}$ - $h^+$) is created. The holes, when react with water, produce hydroxyl radical, which is a strong oxidizing agent. $H_2O$ absorbs the electrons present in the CB to form a superoxide ion, a strong reducing agent. Hence these electron-hole pairs can perform oxidation and reduction reactions simultaneously and convert the pollutants into harmless products which come in contact with the surface of the catalyst [8]. A Schematic representation of the photocatalytic mechanism is shown in Fig. 5.2. The stepwise process is shown below:

$$h\nu \rightarrow e^- + h^+ \quad (5.1)$$

$$h^+ + H_2O \rightarrow H^+ + OH \quad (5.2)$$

$$h^+ + OH^- \rightarrow \cdot OH \quad (5.3)$$

$$e^- + O_2 \rightarrow O_2^- \quad (5.4)$$

$$2e^- + O_2 + 2H^+ \rightarrow H_2O_2 \quad (5.5)$$

$$e^- + H_2O_2 \rightarrow \cdot OH + OH^- \quad (5.6)$$

$$\text{Organic pollutant} + \cdot OH + O_2 \rightarrow CO_2 + H_2O + \text{Other degradation products} \quad (5.7)$$

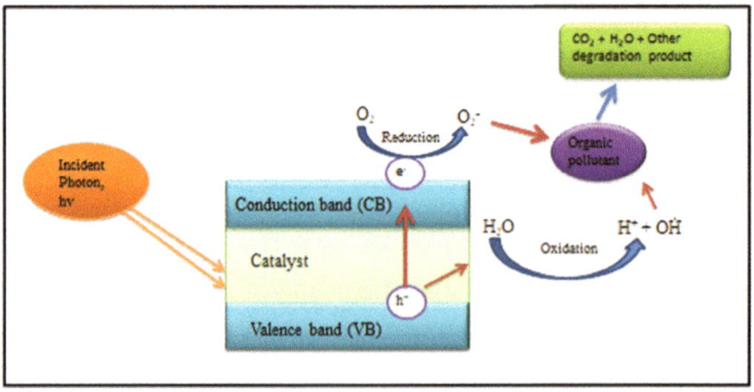


**Figure 5.2** Schematic representation of the photocatalytic mechanism.

### 5.2.1 Mechanism Involving Reactive Oxygen Species

#### 5.2.1.1 Generation of Reactive Oxygen Species

An electron-hole pair was produced as a result of the photoexcitation of a semiconductor catalyst. The suitable wavelength of radiation required for the photoexcitation mainly depends on the band gap energy between the VB and the CB. High energy ultraviolet (UV) radiation is necessary for semiconductor materials with a wide band gap while small band gaps need only lower energy visible light radiation. Semiconductor photocatalysis involves the production of a variety of reactive oxygen species (ROS) and the generation and role of various ROS depend on the reaction conditions and the nature of pollutants involved. In UV irradiated $TiO_2$ photocatalysis, it is believed that hydroxyl radical has a major role in the disintegration of pollutants. Photocatalysis initiated by lower energy visible light does have not much oxidation potential for the generation of hydroxyl radical and therefore singlet oxygen and superoxide anion radical are responsible for the degradation process [9].

#### 5.2.1.2 Role of ROS during Photocatalysis

In semiconductor catalysts, the band edge position determines which ROS is to be produced. Among various ROS, hydroxyl radical is the most efficient oxidizing agent and it undergoes the oxidation processes under UV photoexcitation in aqueous solutions. Hydroxyl radical adds to the multiple bonds or aromatic rings, by hydrogen atom abstraction from CH bonds and electron transfer reaction [10]. Singlet oxygen, hydrogen peroxide, and superoxide anion radical are other oxidants but their oxidation potential is less than the hydroxyl radical. Singlet oxygen is produced by the reduction of superoxide anion radical or photosensitization. Hydrogen peroxide can be formed from the combination of two hydroxyl radicals. The role of $\dot{O}_2^-$ as both oxidizing and reducing agents depends on the reactivity of the pollutant.

## 5.3 Characteristics of Efficient Photocatalysts

Photocatalysis involves redox reactions, i.e., oxidation and reduction which occur simultaneously. So a catalyst is required which can support both processes. A photocatalyst must have the ability to absorb light and undergo photoexcitation that causes the degradation of the reactants used and regenerate itself after each cycle of the reactions. An efficient photocatalyst must be capable of adsorbing reactants on its surface upon suitable energy irradiation ($h\nu \geq E_g$). A suitable photocatalytic system must possess some important features such as desired band gap, high surface area, suitable morphology, stability, reusability, availability, low cost, and chemically and biologically inert nature.

## 5.4 Factors Affecting Photocatalyst Efficiency

### 5.4.1 Light Intensity

In photocatalysis, the degradation rate is directly related to the incident light intensity. The incident light should have energy comparable to or higher than the band gap energy of the catalysts. The activity of photocatalysts is enhanced with an increase in the intensity of the light because the rate of forward reaction (using electron and hole to produce hydrogen and oxygen) is greater than the rate of backward reaction (recombination of electron-hole pairs). Above an optimum intensity, the photocatalytic activity is decreased irrespective of the intensity because high-intensity light promotes the rate of a backward reaction than the forward reaction. For example, $TiO_2$ having a band gap of 3.2 eV is capable of absorbing ultraviolet radiation. When the light intensity was increased from 0 to 20 mW/cm$^2$, the rate of degradation was found to increase. But if the intensity is above 25 mW/cm$^2$ the reaction rate is found to decrease which results in a lower yield of hydrogen [11].

### 5.4.2 Nature of the Photocatalyst

The morphology of the catalyst greatly affects the efficiency of a photocatalyst. Spherically shaped zinc oxide (ZnO) nanoparticles were found to be more effective than rod-shaped ones because of their high surface area. Titanium dioxide ($TiO_2$) can have three different phases: rutile, brookite, and anatase. Among the three phases, the anatase phase is more effective because it has structural stability, high adsorption capability, and the location of the CB. Nanostructured photocatalysts show more degradation activity than their bulk counterparts, materials in nanoscale have a higher surface area which enables different reactants to adsorb on the surface of the catalyst at the same time, and thus reaction rate increases [12].

### 5.4.3 Amount of the Photocatalyst

The reaction rate increases with the amount of the photocatalyst because as the amount of catalyst increases, the radicals responsible for the degradation also increase. But when an optimal limit is reached, a further increase in the amount of catalyst doesn't have any effect on the reaction rate because it makes more complications for the light to penetrate into the medium and makes the catalyst surface inactive [13].

### 5.4.4 Temperature

Photocatalytic activity can be enhanced by increasing the temperature which facilitates the creation of an $e^-$-$h^+$ pair. An increase in temperature results in the desorption of products from the catalyst and it facilitates the production of more electron-hole pairs for further oxidation and reduction which increases the rate of reaction. But when temperature decreases, the desorption rate of products from the catalyst surface also decreases resulting in the reduction in reaction rate. It was reported that the optimum temperature range for the highest reaction rate is 20–80°C. For example, the reaction rate of $TiO_2$ decreased after 80°C due to the recombination of $e^-$-$h^+$ pairs which result in the reduction in the rate of adsorption. [14].

### 5.4.5 pH of the Photocatalytic System

The photoelectron reduces the proton to $H_2$ and therefore the hydrogen ion concentration (pH), plays a major role in the photocatalytic activity. Many researchers reported that the yield of $H_2$ in a basic medium is more than the acidic medium [15]. Nada et al. [16] reported that the yield of $H_2$ in acid medium is more than that in basic medium. They studied the photocatalytic activity of $TiO_2$/$RuO_2$-$MV^{2+}$ catalyst in a methanol-water mixture and explained that in strongly acidic conditions, more and more $H^+$ is adsorbed on the photocatalyst, thus the formation of $H_2$ by the electron is increased, and thus the reaction rate is increased.

## 5.5 Green Nanomaterials as Photocatalysts

### 5.5.1 Metal Nanoparticles

The potential of metal nanoparticles has been exploited in various fields like biodevices, catalysis, sensors, data storage, etc. due to their excellent conductivity, high surface area, and surface plasmonic properties. The characteristic features of metal nanoparticles such as shape, size, etc. play an important role in their photocatalytic activity. The photocatalytic degradation of methyl red dye by silver nanoparticles on UV irradiation was studied by Tamuly et al. [17] and it was observed that the concentration of silver nanoparticles influences the rate of degradation. Alex et al. [18] studied the photocatalytic activity of silver nanoparticles using 0.5 mM MCZ (Mancozeb, an agricultural fungicide) solution (the mechanism of which is shown in Fig. 5.3) and found a broad absorbance peak at 290 nm which shows the degradation of MCZ molecules. The photocatalytic activity of silver nanoparticles (AgNPs) can be either due to the release of $Ag^+$ ions from AgNPs or by the formation of ROS but requires a high amount of energy for the release of $Ag^+$ which cannot be obtained by a halogen lamp and the ozone layer prevents the presence of these high energy UV photons. So, the possible mechanism is the production of ROS.

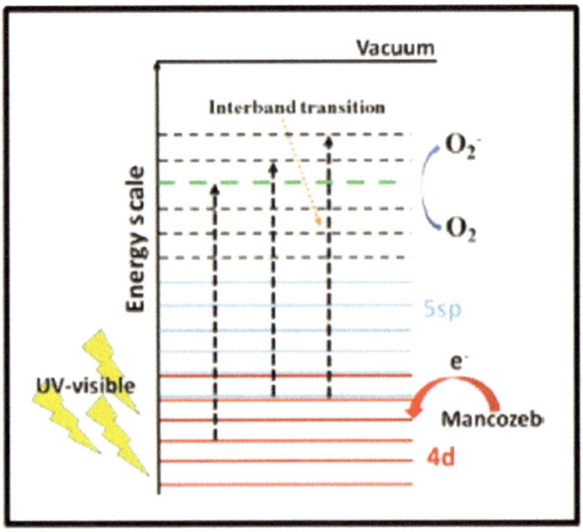


**Figure 5.3** Mechanism of photocatalysis of green synthesized AgNPs. (Reprinted with permission from Alex et al. [18], 2020, American Chemical Society).

### 5.5.2 Metal Oxide Nanoparticles

Metal oxide nanoparticles have great importance in environmental remediation because of their suitable band gap, morphology, and large surface area, and also their ability to produce charge carriers when applied a required amount of energy. Metal oxides fulfill all the requirements for a stable photocatalyst. One of the most important characteristics for an efficient photocatalyst is the band gap which should be in the UV-visible range. Metal oxides are semiconductor catalysts having band gap generally in the range 1.5–3.5 eV, hence they can absorb UV-visible light and are suitable for performing a photocatalytic activity. The important metal oxides which have the said properties are oxides of vanadium, chromium, titanium, zinc, tin, cerium, and copper which can absorb light followed by a charge separation process and form positive holes, that are capable of oxidizing organic substrates [19].

$TiO_2$ is an extensively used photocatalyst because of its high stability and activity, low cost, and non-toxicity. Despite the high activity, the major limitation of the $TiO_2$ catalyst is the rapid recombination of positive and negative charges. It was reported that the enhanced catalytic efficiency of modified $TiO_2$ is because of the reduction in the band gap and decrease in the recombination rate of electrons and holes [20]. Khalafi et al. incorporated stable zinc oxide nanoparticles through microalgae chlorella extract as a reducing agent. The morphological characterization was done using scanning electron microscopy and transmission electron microscopy (TEM) analysis which confirmed spherical morphology with slight agglomeration (Fig. 5.4) Photodegradation of dibenzothiophene was performed using the synthesized ZnO nanoparticles and it was reported that even after five cycles of photodegradation there is little change in the efficiency of the synthesized ZnO nanoparticle catalysts [21]. Iron oxide nanoparticles were also used as photocatalysts for the disintegration of organic pollutants. Among different iron oxides, hematite ($\alpha$-$Fe_2O_3$) having a band gap of 2.1 eV is usually used because of its excellent visible light absorption capacity [22].

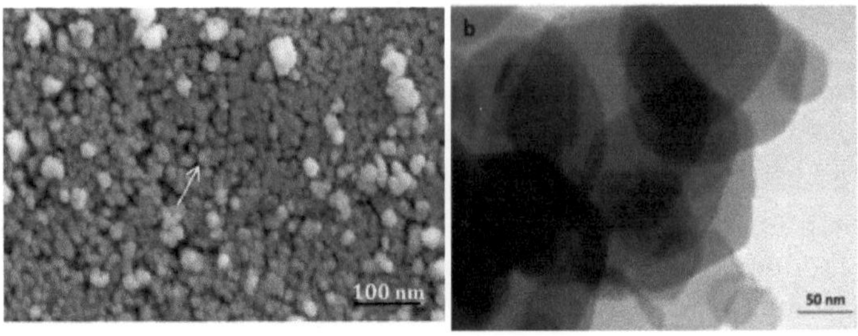


**Figure 5.4** (a) SEM image and (b) TEM image of biosynthesized ZnO nanoparticles using microalgae chlorella extract (Reprinted with permission from Khalafi et al. [21], 2019, by Springer Nature).

### 5.5.3 Metal and Non-metal Doped Metal Oxides

Metal oxide semiconductors have a wide band gap in the range of 2–4 eV and due to this most of the absorptions are in the UV region. To enhance the absorption characteristics and physical properties the metal oxide can be doped with plasmonic noble metals and non-metals. After doping, the parent metal oxide undergoes changes in crystal and electronic structure resulting in changes in the band gap [23]. The commonly used non-metals for doping are N, C, P, Cl, I, S, etc. These dopants can substitute oxygen, metal atomic sites or add to the interstitial sites [24]. Doping creates more dopant sites in between the original band gap, causes new excitations by absorbing visible light radiations. This prevents spontaneous recombination of charge carriers and hence increasing their lifetime. A longer lifetime of charge carriers ensures better activity in photocatalysis.

Metal dopants can also be used for reducing the band gap of metal oxides. Transition metals, group 2 metals (alkaline earth elements), and lanthanides are the normally used metal dopants. In transition metals, first row (3d) metals are widely used as dopants. The 3d orbitals of the transition metals create dopant states in between the initial band gap. In the case of doping of metal oxides, co-doping, tri-doping various elements, or a mixture of metals and non-metals were more productive than mono-doping to strengthen the visible light absorption [25].

ZnO nanoparticles doped with different concentrations of $Eu^{3+}$ in water and other solvents like methanol and ethanol were prepared and studied the catalytic activity against methyl orange. The studies showed that the most effective samples were ZnO:Eu (3%) in water and ZnO:Eu (10%) in methanol [26].

### 5.5.4 Nanowires and Nanorods

Nanorods and nanowires are one-dimensional semiconductor nanostructures studied intensively for industrial applications. The photocatalytic activity of $Fe_2O_3/Bi_2S_3$ nanorods synthesized via a one-step hydrothermal route against organic pollutants such as methylene blue and phenol was reported [27]. They have shown that the maximum of 90% degradation of methylene blue was recorded after 300 min under visible radiation using 0.6 M $Fe_2O_3/Bi_2S_3$. Rutile $TiO_2$ nanorods can be formulated using a hydrothermal method by changing the time and concentration of glycolic acid (GA) (Fig. 5.5a) [28]. The shape of the synthesized nanorods was found to be independent of the GA concentrations as shown in the TEM images (Fig. 5.5b). They also studied the impact of calcination temperature on the properties of the synthesized $TiO_2$. After calcination, they were found to become thick, deform into sub-sphere and particle size becomes larger. They reported that large particle size causes a wide separation of redox sites and this decreases the recombination rate of the photoexcited electrons and holes thus enhancing the photocatalytic activity. The photocatalytic efficiency of $TiO_2$ nanorods can be increased by Pt-loading because Pt decreases the recombination of charges by storing the photo-generated electrons [28]. Nanorods possess a rather long lifetime of photogenerated electrons due to the smooth electron movement in 1D morphology. Anyhow the recombination takes place effortlessly on the surface of nanorods resulting in the reduction of photocatalytic activity.

ZnO nanowires can be used as photocatalysts to inactivate bacteria and viruses, for the degradation of pollutants such as dyes, pesticides, etc. under suitable light irradiation [30]. To enhance the semiconductor properties of nanowires, a doping strategy could be made use of which can control the properties such as band gap, electrical conductivity, and ferromagnetism. Metal doped with nanowires includes Co, Cu, Ni, Al, Eu [31], etc., and non-metals such as C, N, Cl, P [32]. Nanowires codoped with two different elements include Mn and Co, Mn and Li, and Li and N [33].

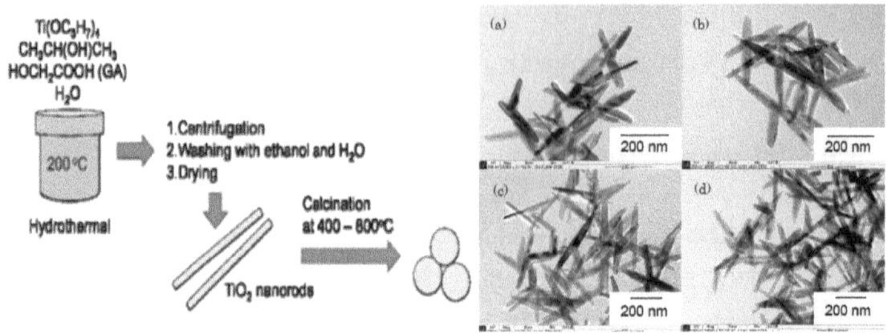


**Figure 5.5** (a) Synthesis of $TiO_2$ nanorods and $TiO_2$-*X* (*X* = 400–800), (b) TEM images of $TiO_2$ nanorods synthesized with glycolic acid concentration of (a) 0.8, (b)1.2, (c) 1.6, and (d) 2.0 mol $L^{-1}$. (Reprinted with permission from Yamazaki et al. [28], 2018, American Chemical Society).

## 5.5.5 Quantum Dots

Quantum dots (QDs) are 0-dimensional nanostructures with an aspect ratio near unity. QDs exist in nanoscale regime showing fluorescence emissions with different colors which are found to be size-dependent. Among these materials, the most important one is carbon dots. In carbon dots, photoexcitation occurs because of the π-plasmon absorption of the nanoparticles. It can strongly absorb the bulk of the solar spectrum including UV-visible and near-infrared radiations.

Generally, carbon dots are formulated by the process of carbonization of carbon-rich precursors during one-pot synthesis. Carbon dots were synthesized from chitosan via hydrothermal carbonization [34]. Choi et al. demonstrated the redox nature of photoexcited carbon dots [35]. The photoirradiation of the solution of carbon dots with metal (Ag, Au, and Pt) salts causes the deposition of metal on carbon dots. These metals deposited on the surface acting as a strong electrophile draws electrons from carbon dots preventing the recombination resulting in effective quenching of fluorescence emissions. Also, these deposited metals can act as co-catalysts and the carbon dots could be used as efficient photocatalysts.

## 5.5.6 Biopolymeric Nanomaterials

Biopolymers are natural polymers made up of monomeric units, covalently binding to form large-sized molecules. Materials that are synthesized from biological sources such as vegetable oils, fats, proteins, amino acids, resins, sucrose, etc. are biopolymers. On the basis of degradability, polymers are classified as biodegradable and non-biodegradable. The basis on the structure, biopolymers are generally grouped into polyesters, polysaccharides, polyamides, polycarbonates, etc. Cellulose, nano cellulose, chitin, lignin, starch, etc. are the most extensively studied biopolymers for photocatalytic activity. Nano cellulose in the form of nanocrystals or nanofibers with sulphonyl or carboxylic acid functionality has high activity for cationic dyes like methylene blue and malachite green while amino-functionalized show higher activity for anionic dyes. Cu-doped cellulose nanofibers/nano titanium dioxide was synthesized by hydrolysis–precipitation method and used to decolorize organic dyes [36]. They also reported that the prepared photocatalyst can be recycled. Chitosan-silver nanocomposite was synthesized and evaluated the photocatalytic activity [37]. They showed a maximum of 81% decolorization of methyl orange under visible light irradiation.

## 5.6 Conclusions and Future Outlook

The major issue related to rapid urbanization is the release of various toxic pollutants directly or indirectly into the surroundings. For the past few years, several studies have been focused on the use of green nanomaterials as photocatalysts due to their outstanding physicochemical properties and now they are advised as the best choice to keep the environment safe, mainly in the photocatalysis process to degrade the harmful toxic pollutants. The success of photocatalytic dye degradation using various green nanomaterials is attributed to the green process, uniform dispersion of nanoparticles, non-toxic and recyclable nature. The photocatalytic efficiency of various green nanomaterials could be made use of in the conversion of toxic pollutants into harmless byproducts and thus effective environmental remediation. However, the potential exploration of novel green nanotechnology is truly challenging in the sense that the risk of implementing such an innovative strategy compared with conventional technologies needs to be considered seriously and a more rational approach for the design of suitable green nanomaterials which satisfy all the quality requirements is highly necessary for the coming years.

## References


1. Danish, M.S.S., Estrella, L.L., Alemaida, I.M.A., Lisin, A., Moiseev, N., Ahmadi, M., Nazari, M., Wali, M., Zaheb, H., Senjyu, T., Photocatalytic Applications of Metal Oxides for Sustainable Environmental Remediation, Metals, 2021. 11: pp. 80.
2. Kalita, E., Baruah, J., Thomas, S., Tresa Sunny, A., Velayudhan, P., Environmental Remediation. In Colloidal Metal Oxide Nanoparticles, Eds.; Elsevier: Amsterdam, the Netherlands, 2020. pp. 525–576.
3. Englande, A.J., Krenkel, P., Shamas, J., Eds., Wastewater Treatment and Water Reclamation. In Reference Module in Earth Systems and Environmental Sciences; Elsevier: Amsterdam, The Netherlands, 2015.
4. Zaharia, C., Suteu, D., Muresan, A., Muresan, R., Popescu, A., Textile Wastewater Treatment by Homogeneous Oxidation with Hydrogen Peroxide, Environ. Eng. Manag. J, 2009. 8: pp. 1359–1369.
5. Gusain, R., Gupta, K., Joshi, P., Khatri, O.P., Adsorptive Removal and Photocatalytic Degradation of Organic Pollutants Using Metal Oxides and their Composites: A Comprehensive Review, Adv. Colloid Interface Sci., 2019. 272: pp. 102009.
6. Gautam, S., Agrawal, H., Thakur, M., Akbari, A., Sharda, H., Kaur, R., Amini, M., Metal Oxides and Metal-Organic Frameworks for the Photocatalytic Degradation: A Review, J. Environ. Chem. Eng., 2020. 8: pp. 103726.
7. Hitam, C.N.C., Jalil, A.A., A Review on Exploration of $Fe_2O_3$ Photocatalyst towards Degradation of Dyes and Organic Contaminants, J. Environ. Manag., 2020. 258: pp. 110050.
8. Schwanke, A.J., Balzer, R., Pergher, S., Martínez, L., Kharissova, O., Kharisov, B., Eds., Microporous and Mesoporous Materials from Natural and Inexpensive Sources. In Handbook of Ecomaterials; Springer: Berlin/Heidelberg, Germany, 2019. pp. 3379–3399.
9. Theerthagiri, J., et al. Recent Developments of Metal Oxide Based Heterostructures for Photocatalytic Applications towards Environmental Remediation, J. Solid State Chem., 2018. 267: pp. 35–52.
10. Buxton, G. V., Greenstock, C. L., Helman, W. P., Ross, A. B., J. Phys. Chem. Ref. Data, 1988. 17. pp. 513–886.
11. Al-Hamdi, A.M., Rinner, U., Sillanpää, M., Tin Dioxide as a Photocatalyst for Water Treatment: A review, Process Saf. Environ. Prot., 2017. 107: pp. 190–205.

12. Magdalane, C.M., et al., Photocatalytic Decomposition Effect of Erbium-doped Cerium Oxide Nanostructures Driven by Visible Light Irradiation: Investigation of Cytotoxicity, Antibacterial Growth Inhibition Using Catalyst, J. Photochem. Photobiol. B Biol., 2018. 185: pp. 275–282.

13. Saikia, L., et al., Photocatalytic Performance of ZnO Nanomaterials for Self-sensitized Degradation of Malachite Green Dye under Solar Light. Appl. Catal. A Gen., 2015. 490: pp. 42–49.

14. Dong, F., et al., Enhancement of the Visible Light Photocatalytic Activity of C-doped $Tio_2$ Nanomaterials Prepared by a Green Synthetic Approach. J. Phys. Chem. C, 2011. 115: pp. 13285–13292.

15. Liu, S., et al., Multifunctional $TiO_2$ Over Layers for p-Si/n-CdS Heterojunction Photocathode with Improved Efficiency and Stability. Nano Energy, 2018. 53: pp 125–129.

16. Nada, A., et al., Enhancement of Photocatalytic Hydrogen Production Rate Using Photosensitized $TiO_2/RuO_2$-MV2þ. Int. J. Hydrogen Energy, 2008. 33: pp. 3264–3269.

17. Tamuly, C., Hazarika, M., Bordoloi, M., Das, M. R., Photocatalytic Activity of Ag Nanoparticles synthesized by using Piper Pedicellatum C.DC fruits. Mater. Lett, 2013. 102–103: pp. 1–4.

18. Alex, K. V., Pavai, P. T., Rugmini, R., Prasad, M. S., Kamakshi, K., Sekhar, K. C., Green Synthesized Ag Nanoparticles for Bio-Sensing and Photocatalytic Applications, ACS Omega, 2020. 5: pp. 13123–13129

19. Pelizzetti, E., Minero, C., Metal oxides as Photocatalysts for Environmental Detoxification, Comments Inorg. Chem., 1994. 15: pp. 297–337.

20. Li, R., Li, T., Zhou, Q., Impact of Titanium Dioxide ($TiO_2$) Modification on its Application to Pollution Treatment—A Review, Catalysts, 2020. 10: pp. 804.

21. Khalafi, T., Buazar, F., Ghanemi, K., Phycosynthesis and Enhanced Photocatalytic Activity of Zinc Oxide Nanoparticles toward Organosulfur Pollutants, Sci. Rep., 2019. 9: pp. 6866.

22. Xu, P., Zeng, G. M., Huang, D. L., Feng, C. L., Hu, S., Zhao, M. H., Lai, C., Wei, Z., Huang, C., Xie, G. X., Liu, Z. F., Sci. Total Environ., 2012. 424: pp. 1–10.

23. Liu, Y., Wang, W., Xu, X., Marcel Veder, J. -P., Shao, Z., Recent Advances in Anion-Doped Metal Oxides for Catalytic Applications, J. Mater. Chem. A, 2019. 7: pp. 7280–7300.

24. Chen, H., Dawson, J. A., Nature of Nitrogen-Doped Anatase $TiO_2$ and the Origin of Its Visible-Light Activity, J. Phys. Chem. C, 2015. 119: pp. 15890–15895.

25. Zhang, Z., Wu, Q., Johnson, G., Ye, Y., Li, X., Li, N., Cui, M., Lee, J. D., Liu, C., Zhao, S., Li, S., Orlov, A., Murray, C. B., Zhang, X., Gunnoe, T. B., Su, D., Zhang, S., Generalized Synthetic Strategy for Transition-Metal-Doped Brookite-Phase TiO2 Nanorods. J. Am. Chem. Soc, 2019. 141: pp.16548–16552.

26. Trandafilovic, L. V., Jovanovic, D. J., Zhang, X., Ptasińaninska, S., DramićM.D., Enhanced Photocatalytic Degradation of Methylene Blue and Methyl Orange by ZnO: Eu nanoparticles, Appl. Catal. B, 2017. 203: pp. 740–752.

27. Helal, A., Harraz, F.A., Ismail, A.A., Sami, T.M., Ibrahim, I.A., Hydrothermal Synthesis of Novel Heterostructured $Fe_2O_3/Bi_2S_3$ Nanorods with Enhanced Photocatalytic Activity under Visible Light, Appl. Catal. B, 2017. 213: pp. 18–27.

28. Yamazaki, Y., Azami, K., Katoh R., Yamazaki S., Developing Active $TiO_2$ Nanorods by Examining the Influence of Morphological Changes from Nanorods to Nanoparticles on Photocatalytic Activity, ACS Appl. Nano Mater., 2018.

29. Tamaki, Y., Hara, K., Katoh R., Tachiya, M., Furube, A., Femtosecond Visible-to-IR Spectroscopy of $TiO_2$ Nanocrystalline Films: Elucidation of the Electron Mobility before Deep Trapping, J. Phys. Chem. C, 2009. 113: pp. 11741–11746.

30. Wang, X., Wang, W., Liu, P., Wang, P., Zhang, L., Photocatalytic degradation of E. coli membrane cell in the presence of ZnO nanowires, J Wuhan Univ. Technol. Mater. Sci. Ed., 2011. 26: pp. 222–225.

31. Baiqi, W., et al., Photoluminescence properties of Co-doped ZnO nanorods array fabricated by the solution method, Phys. E, 2009. 41: pp. 413–417.

32. Dai, Z., et al., C-doped ZnO Nanowires: Electronic Structures, Magnetic Properties, and a Possible Spintronic Device, J Chem. Phys., 2011. 134.

33. Das, S. N., Choi, J. H., Kar, J. P., Lee, T. I., Myoung, J. M., Fabrication of p-type ZnO Nanowires based Heterojunction Diode, Mater. Chem. Phys., 2010. 121: pp. 472–476.

34. Yang, Y., Cui, J., Zheng, M., Hu, C., Tan, S., Xiao, Y., Yang, Q., Liu, Y., One-Step Synthesis of Amino-Functionalized Fluorescent Carbon Nanoparticles by Hydrothermal Carbonization of Chitosan, Chem. Commun, 2012. 48: pp. 380–382.

35. Choi, H., Ko, S. -J., Choi, Y., Joo, P., Kim, T., Lee, B. R., Jung, J. -W., Choi, H. J., Cha, M., Jeong, J. -R., Hwang, I. -W., Song, M. H., Kim, B. -S., Kim, J. Y., Versatile Surface Plasmon Resonance of Carbon-Dot-Supported Silver Nanoparticles in Polymer Optoelectronic Devices, Nat. Photon., 2013. 7: pp. 732–738.
36. Li, Y., Pan, Y., Zhang, B., Liu, R., Adsorption and Photocatalytic Activity of Cu-doped Cellulose Nanofibers/nano-titanium Dioxide for Different Types of Dyes, Water Sci. Technol., 2020. 82: pp. 1665–1675.
37. Nithya, A., Jothivenkatachalam, K., A Versatile Effect of Chitosan-silver Nanocomposite for Surface Plasmonic Photocatalytic and Antibacterial Activity, J. Photochem. Photobiol. B Biol., 2015. 153: pp. 412–422.

# Chapter 6

# Green Nanomaterials for Wastewater Treatment Analysis

**Uyiosa Osagie Aigbe,[a] Kingsley Eghonghon Ukhurebor,[b] Robert Birundu Onyancha,[c] Benedict Okundaye,[d] and Otolorin Adelaja Osibote[a]**

[a]*Department of Mathematics and Physics, Faculty of Applied Sciences, Cape Peninsula University of Technology, Cape Town, South Africa*
[b]*Department of Physics, Faculty of Science, Edo State University Uzairue, Edo State, Nigeria*
[c]*Department of Physics and Space Science, School of Physical Sciences and Technology, Technical University of Kenya, Nairobi, Kenya*
[d]*School of Art, Design, and Architecture, University of Huddersfield, Queensgate Huddersfield, UK*
uyi4we@gmail.com

The condition of water globally is becoming more thought-provoking owing to the expanding population and change in the climate. With a continuous increase in the magnitude of heavy metals (HM) contamination in water and wastewater, there is a greater need for new advanced water technologies that would guarantee a safe and clean water supply. Hence, the main focus of this chapter is centered on the emerging nano sorbent for water and wastewater treatment

*Green Nanoarchitectonics*
Edited by Kaushik Pal

ISBN 978-981-4968-50-8 (Hardcover), 978-1-003-31860-6 (eBook)
www.jennystanford.com

that use diverse sources of plant, microorganisms, and biocompatible green reagents for the production of green nanomaterials (GNMs) utilized for the decontamination and recovery of HM from industrial effluents due to their practicality, minimal cost and eco-friendly drawn from existing publications. From the review of existing publications, it was observed that the sorption of most HM was found to be pH-dependent, with optimum elimination detected in the pH range of 2–10. Also, the sorption process of most HM was superlatively defined employing the Langmuir (LGR) and pseudo-second-order (PSO) models and the optimum sorption capacities were in a range of using GNMs. Conclusively, recommendations in the form of the future outlook for the integration of GNMs for HM ions regarding environmental sustainability are highlighted.

## 6.1 Ecological Survey of the Global Impact of GNMs

Society and nations advancement is ascertained and this occurs with serious ecological issues (Aigbe, et al., 2018). With regards to the quantity and quality of water, its scarcity characterizes a serious threat to the present and future well-being of individuals globally, most especially to developing nations' populace. Ecological water administration is a critical aspect of resolving poverty, parity, and associated issues. A global population of 83% (636 million) is deprived of access to an improved source of drinking water in rustic regions. Also, the number of individuals with no access to reliable water for drinking is two to three times above average than the official assessments (Hashem, 2014).

Accessible water supplies are reducing owing to low precipitation, growth in the human population, more severe health-based procedures, and challenging needs from an array of users. While, drinking water of good quality in current years has been a huge challenge due to the enhanced contamination of water bodies (Nazal, 2019, and Aigbe et al., 2019).

Heavy metal ions (HMI) are a category of inorganic pollutants that are introduced into water resources through unprocessed waste discharges mostly from fertilizers (from agricultural activities), mining (from mining activities), batteries (from both industrial and domestic activities), pesticides (from agricultural activities), refining (from industrial activities), tanneries (from tanneries activities) as well as the paper industries. The pathway for HMI exposure is predominantly through ingestion, dermal contact, and inhalation (Onyancha, et al., 2021). In various and specific environmental situations, HMI release happens naturally. Such release comprises of spraying of sea salt, eruptions in volcanic processes, weathering of rock, forest fire, wind-borne soil constituent part, and biogenic sources. In most developing nations, HMI occurrence in water resources is due to anthropogenic emission and natural processes (Aigbe and Osibote, 2020).

Contamination of water and its remediation is a global and growing issue that has become evident. HM (loids) presence and maintaining its level in water are of particular concern due to their toxicity and carcinogenicity, which may trigger serious damage to a variety of humanoid body systems (Onyancha, et al., 2021). High concentrations (CCTN) of HMI like mercury (Hg), nickel (Ni), chromium (Cr), cadmium (Cd), and lead (Pb) occurrence in natural water resources and industrial wastewater streams are critical to human health and the environment owing to their huge noxiousness and bioaccumulation via food chain has being a huge issue in developed and developing nations (Aigbe, et al., 2018).

Worldwide, the water need for industrial processes has increased enormously, but its abuses and insufficient appropriate treatment consistently led to pollution of water resulting in its scarcity. To stun this challenge, various water treatment and purification methods for different industrial processes have been employed by industries (Ukhurebor, et al., 2021). Hence, there is a critical need for pollutants removal from wastewater to make it eco-friendly (Jawed, Saxena, and Pandey, 2020).

The various established methods employed to solve problems relating to water contamination with HMI include chemical precipitation (Onyancha, et al., 2021), sorption (Aigbe, et al., 2018), membrane filtration (Barman, et al., 2021), photocatalysis (Abdel-Aziz, Farag, and Abdel-Gawad, 2019), ion exchange (Sharma, et al., 2017), reverse osmosis (Wenten, 2016), etc. But for a good quality sequestration outcome, a common situation is the combination of different techniques for the management of adulterated water among the listed techniques (Onyancha, et al., 2021).

Among the various methods employed for HMI decontamination, sorption is desirable in terms of design, operation and scale-up, elevated capacity, improved regeneration, and minimal cost. Also, the quality of waste produced is far better than the rest of the processes due to the high affinity of these sorbents for metal ions (MI). This affinity may be due to electrostatic interactions of the solute with the surface of the sorbent (SS), Van der Waals attraction, or chemisorptions (Gangadhar, Maheshwari, and Gupta, 2012). This flexible and low-cost technique works effectively to decontaminate wastewater with a pruned CCTN of metals wastes and the efficiency of this technique depends on the sorbents class (Karman, Diah, and Gebeshuber, 2015).

Sorbents are generally porous materials with a huge surface area owing to their elevated porosity. Due to this, the interior part is mostly larger than the exterior part. Sorbents can be classified into natural inorganic (like clays as well as other minerals such as calcite, goethite, hydroxy-apatite, calcareous soils, fly ash, zeolite, slags, sludges, and modified asphaltite), natural organic (such as sawdust, shells of coconut, waste from corncob, waste from tea, rice undersides, bark, shells of hazelnut, chitosan, wool, and peat) and man-made sorbents (nano-sized MI oxides, zero-valent iron, improved NMs) (Pfeifer and Škerget, 2020). While traditional sorbents can confiscate HMI from waste discharges, their minimal productivities, creation of excessive sludge, and minimal sorption capacities restrict their use. Presently, NMs are employed as new sorbents for the sorption of HMI from waste discharges to spurn the problems linked with traditional sorbents (Onyancha, et al., 2021).

A research area that has advanced at a very rapid rate and which presents various desirable alternatives for the management of water and wastewater is nanotechnology (NT) in relation to nanoscience. NM has gained consideration for the degrading and remedying of noxious organic and inorganic contaminants into non-noxious free resources due to their distinctive physicochemical features like their huge catalytic activity, high chemical/physical and thermal permanence, huge individual surface area, improved chemical reactivity, and sturdy ability to transmit electron. This has inspired the design of appropriate bioinspired biogenic and greener approaches that are environmentally responsive, secure, and cost-efficient for innovative and effective advancement of nanoscale sorbents and catalysts for water pollutants removal and degrading (Nasrollahzadeh, et al., 2021 and Al-Qahtani, 2017). A vast majority of these NMs are compatible with current treatment techniques and can be infused into an active system. New engineered NMs' huge surface areas and extraordinary reactivity encourage their use for the sequestration of harmful pollutants and in this perspective, there has been considerable interest in the green procedure development for the removal of ionic metal species (Nasrollahzadeh, et al., 2021). The modern age of green synthesis techniques has attracted huge interest in existing studies and advancements in materials science and technology. The production of NMs employing extract of plants via green synthesis has fascinated much systematic attention owing to its operational simplicity and minimal cost (Singh, et al., 2018 and Weng, et al., 2018).

The objective of green NT (GNT) development, that employs biological routes for the production of NMs is the reduction in harmful materials production. Also, the amount of energy involved in GNT is relatively low in comparison to other technologies as well as practically minimal noxious chemicals are created for the duration of the synthesis and their eco-friendliness and compatibility are mostly high (Bahrulolum, et al. 2021). Intrinsically, the number of research conducted from 2012–2021 using green NMs for HMI removal has increased as noticed in Fig. 6.1 which is a histogram for the number of publications containing the keyword "green-NMs (GNMs)" for HMI removal" adopted from SCOPUS. This book chapter will look at existing developments and potential chances of utilizing green synthesized NMs for the treatment of water and wastewater under varying conditions like pH, sorbent dosage (SD), initial metal CCTN, and temperature. Also discussed are the mechanism of sorption using GNMs for HMI sorption employing isotherm (IST) and kinetic (KNT) models and thermodynamic (TMD) factors.

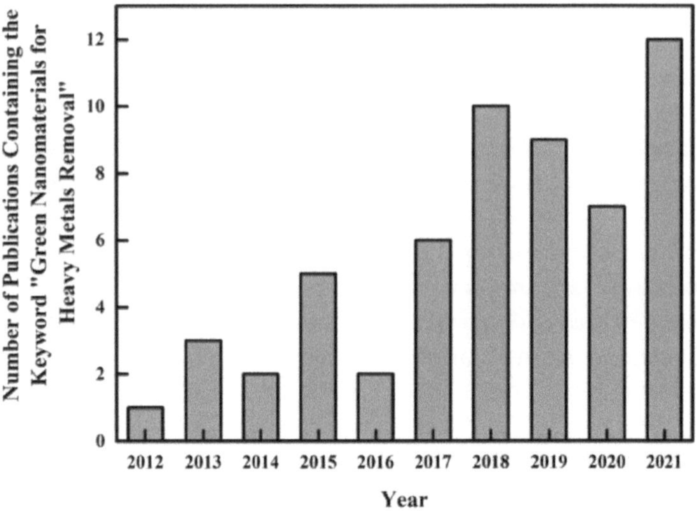


**Figure 6.1** Histograms for the number of publications containing the keyword "GNMs for HMI removal" as adopted from SCOPUS (Note: 2021 is from January–July).

## 6.2 HMI Removal Employing GNMs

The green chemistry (GC) concept started in 1998 through the introduction of uncontaminated, non-noxious, and ecologically responsive reducing and overlaying agents such as sugars, alkaloids, amino acids, carbonyl, polyphenols, terpenoids, protein (PT), and polysaccharides from florae or plants parts as an alternative to extremely destructive chemicals employed for NPs production. Furthermore, the GC aim is to employ bio-based properties under ambient atmospheric temperature and atmospheric pressure by decreasing experimental and ecological hazards. In GNT, the choice of green alternatives for reducing, covering, and stabilizing agents are the key crucial influences that set off clean synthesis procedures that are eco-friendly (Ali, 2017).

GNT enhances sustainability by utilizing NT products and devices. It also involves the creation of green nano-products (NP) and employing these products in support of sustainability and improving the ecosystem. The two dimensions of GNT include the first dimension NP (providing solutions to ecosystem problems such as contaminants prevention, harmful waste site remediation, contaminants streams clean up and desalinate water) and the second dimension NP comprises of the NMs production, which can support in reducing impairment to humanoid health or the ecosystem (Hashem, 2014).

GNMs are created through biological techniques and during their synthesis, organic materials like florae, microbes, and organic polymers such as carbohydrates (CHD), lipids, and PT are vigorously employed. Owing to the specific diminutive size of the NMs, they play a great part in the treatment of wastewater and give opportunities for specific points targeting. Various forms of biosynthesized NMs, which are employed for water contamination control are phytogenic magnetic NPs, zinc oxide NPs, biopolymer-layered metal NPs, silver (Ag)-infused cyclodextrin nanocomposites, Ag in Aloe vera plants extract as well as the photocatalytic titania NPs. NMs offering an optional approach for contaminants removal from water bodies are green NPs, as they provide an economic, appropriate, and environmentally friendly way for the treatment of wastewater (Singh, Yadav, and Kuddus, 2020). Employing biology-based remediation processes is desirable owing to their low cost of operation than physiochemical processes. The utilization of biological-based materials for sorption processes holds huge potential as they are extremely cost-efficient and naturally benevolent (Karman, Diah, and Gebeshuber, 2015).

Microwave and ultrasound are the various environmentally responsive and biological methods for the biogenic synthesis of NMs employing florae and microbes such as bacteria, yeast, viruses, algae, fungi, and waste constituents or the combination of the biogenic techniques with different instigation methods. The historic use of creatures in the synthesis of bio-NMs started in the 1980s when Beveridge and Murray assess the gold (Au)-NPs synthesis by employing Bacillus subtilis as an aerobic, gram-positive bacterium. Microbes can sorb and amass MI to their elemental arrangement, by secreting a huge number of enzymes via cell movements and the microbial creation of NPs are subjected to the reductive enzymes/metabolites on the walls of the cell, their position on the cell, or soluble enzymes discharge. Plant polyphenols, microorganisms, algae, enzymes, and industrial or agricultural wastes are utilized in biological/biogenic methods, and among these NMs, enzymes and their metabolites such as PT, polysaccharides, peptide chains, CHD, and nucleic acids are employed as capping or reducing agents for MI oxides to produce various NPs and functionalized NPs on inorganic supports (Nasrollahzadeh, et al., 2021).

Microorganisms and florae have all it takes to alter metal oxidation conditions under perfectly specified ecological requirements. Due to these entities (viruses, bacteria, yeast, fungi, algal, and florae) having the ability to biochemically process where NPs formations arise owing to the oxidation-reduction of MI via enzymes, sugars, PT, polyphenol, aldehyde, carboxyl groups by the species of microbes and plants to form phytogenic magnetic nanoparticles of FeO, $Fe_3O_4$, and $Fe_2O_3$. Various types of biogenic NPs (BNPs) are produced employing numerous MIs employing various biological samples such as Au, Ag, alloy, silica, titanium, palladium, selenium, antimony, sulfide, etc. These BNPs have been utilized for an extensive array of applications in ecological protection and biomedical (Ali, et al., 2017).

For microorganisms like algae, viruses, bacteria, and fungi are regarded as biofactories due to NPs being created extracellularly/intracellularly and the production of these metal NPs by bacteria is generally because of the self-shielding machinery of the cell of bacteria en route for MI. Through the change of the oxidation state of the MI, noxious foreign MI incursion into the cell results in bioaccumulation/biomineralization. While in fungi, NPs are created via biomineralization also known as myco-synthesis and their elevated content of PT and exceptional metal tolerance capacity make them more valuable than bacteria. While metal NPs are produced using viruses as bio-templates and in algae, the bioactive materials aid to decrease MI and stabilize them in their medium (Puthukkara, Jose, and Lal, 2021).

Green NPs generally utilized for the management of water and wastewater systems are metal oxide NPs like $TiO_2$, ZnO, $Fe_3O_4$, and Ag NPs (Singh, Yadav, and Kuddus, 2020). Fig. 6.2 shows the various classes of GNMs, the synthesis methods employed for their production, their environmental applications, and their derivatives.

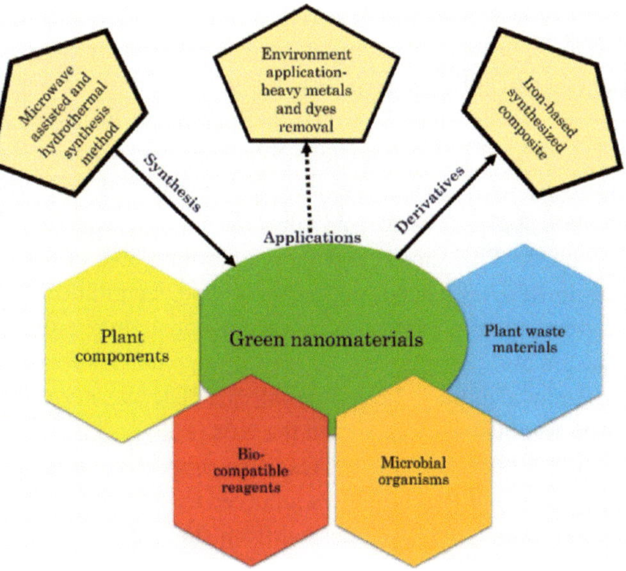


**Figure 6.2** Representations of the various classes of GNMs, the synthesis methods employed for their production, their environmental applications, and their derivatives.

In the study by Das, et al., 2020, magnetic NPs (MNPs) were created using an inexpensive and environmentally friendly approach in an open-air environment employing crude latex of *Jatropha Curcas* (JC) and leaf extract of *Cinnamomum Tamala* (CT). The synthesized MNPs ($Fe_3O_4$) were utilized in the treatment of waste effluents with bacteria, HMI, and dye removal. The sorption of methyl blue (MB) dye to JC-$Fe_3O_4$ was superlatively depicted employing the Freundlich (FRH) as well as the PSO models. The maximum sorption capacity (SNC) as assessed from the LGR model using JC-$Fe_3O_4$ was 466.6 mg/g. Employing both NPs for the elimination of Cu (II) and Co (II), the LGR model was determined to best depict the sorption of these HMI to the NPs, with maximum SNC of 513.7 and 463.23 mg/g for Co (II) and Cu (II) employing CT-$Fe_3O_4$. While the corresponding maximum SNC using JC-$Fe_3O_4$ for Co (II) and Cu (II) were 501.3 and 543.3 mg/g. Also, employing both NPs for antibacterial activity, it was found that the bacterial colony in the case of pond water decreased by more than 50% when compared with untreated pond water.

Zero valent Ag NPs were created using an eco-friendly technique or a green approach was employed as a sorbent for the treatment of wastewater with Cd (II). The Ag NPs were synthesized from the extract of the Ficus tree (*Ficus Benjamina*) leaf. The removal percentage of Cd (II) using these NPs improved with a rise in the solution pH, with a peak removal percentage of 75.5 and 85% determined at pH 6. The sorption process of Cd (II) was superlatively defined by the FRH model and with a resolute SNC of 0.845 mg/g. Iron oxide NP (IONP) created using the co-precipitation technique and capped by the extract of tangerine peel to prevent particle aggregation and decrease the widths of the particles was employed for the removal of Cd (II) (Al-Qahtani, 2017). It was noticed that the optimum removal of Cd (II) onto the NPs (90%) transpired at pH 4 using an SD of 0.4 g/100 ml. It was also reported in this study that the sorption of Cd (II) to the NPs followed the FRH and PSO models. The determined SNC was assessed from the LGR model to be 15.5 mg/g (Ehrampoush, et al., 2015).

While Lingamdinne, Koduru, and Rao Karri, 2019, study described the green synthesis of IONP via the co-precipitation technique, which was modified with extracts of tangerine peels (T-$Fe_3O_4$) and employed in the sequestration of Pb (II) from a water-soluble solution. 99% of sorbate was eliminated from the water-soluble solution using 0.6 g/L of the sorbent at pH 5.0. The sorption of Pb (II) to the bio-NMs was ideally described using the LGR and PSO models and the determined SNC was evaluated to be 101.2 mg/g at room temperature. It was superficial that the amount of MI sorbed to the sorbent specific area was improved with temperature rise and the sorption process was endothermic as well as being spontaneous. A significant amount of Cr (VI) ion under varying factors was observed to be decontaminated from an aqueous environment using an innovative MXene-CS composite synthesized utilizing the solution synthesis technique. The maximum SNC of Cr (VI) confiscated was 22.3 mg/g at pH 4.0. This was attributed to the electrostatic interface between the positive (+ve) charged sorbent and the negative (–ve) charged $Cr_2O_7^{2-}$. While the sorption of Cr (VI) to the composite was best designated by the LGR and PSO models, with a determined SNC of 43.1 mg/g. The TMD factors show that the process of sorption was spontaneous and endothermic (Wan, et al., 2021).

A fusion of condensed graphene oxide (rGO) and iron NPs (rGO/FeNPs) was created through the in-situ GO and $FeCl_3$ reduction using extract of Eucalyptus leaf in one pot. The synthesized rGO/FeNPs were employed in the removal of Pb (II) ions, with maximum removal of 82.4% obtained. Also, a comparatively extraordinary selectivity for Pb (II) ions was shown by rGO/FeNPs in the competitive sorption with Ca and Mg cations (Xiao, et al., 2019).

FeNPs effectively created from the extract of Eucalyptus leaf via the green technique were utilized for Cr (VI) removal under varying experimental factors. The elementary composition of the created FeNPs surface from the BET result shows a composition of carbon (C: 41.20 weight% [wt%]), oxygen (O: 18.30 wt%), and Fe (40.30 wt%). C and O were ascribed to the polyphenol group and other C, O encompassing particles presence. After sorption of Cr (VI), a significant amount of Cr (1.90 wt%) was noticed on the FeNPs surface. The removal efficiency of 98.90% was observed using a pH of 4.2, which subsequently decrease with an increase in solution pH to 94.70 and 97.80% for solution pH of 7.0 and 10.0, correspondingly. The sorption process of Cr (VI) was ideally depicted by the LGR and PSO models and the determined SNC was 20.5 mg/g. It was also discovered that FeNPs can be reused up to four regeneration-desorption cycles for the elimination of Cr (VI) with elimination efficiencies of 98.60, 92.40, 63.10, and 55.80%, respectively. This showed the exceptional activity and stability of the created NPs (Jin, et al., 2018).

In the study by Mahanty et al., (2020), *manglicolous fungi Aspergillus tubingensis* (STSP 25) separated from samples of rhizospheric sediment of *Avicennia officinalis* was used to bio produce a new myco-synthesis of IONP and was utilized for the confiscation of various HMI (Pb (II), Ni (II), Cu (II) and Zn (II)) from water. The peak removal for all metals was found at pH 6, which was attributed to the electrostatic interface between the cationic HMI and the IONPs. HMI sorption followed the LGR and PSO models and the optimum sorption capacities were 9.206, 9.666, 8.335, and 9.106 mg/g for Pb (II), Ni (II), Cu (II), and Zn (II). From the determined TMD factors, the sorption process of all MI followed spontaneous chemisorption which proceeded endothermically. While a five sorption/desorption cycle of high renewal capacity of the sorbent was found using this sorbent for HMI removal. Table 6.1 gives a brief of the various GNMs employed for HMI removal, with the determined sorption capacities, optimum pH, dosage, IST and KNT models and TMD factors also included in Table 6.1.

**Table 6.1** Outlines of various green NMs employed for the removal of HMI

| Green NMs | HMI | pH | IST | KNT | Maximum SNC (mg/g) | Dosage | TMD factors | References |
|---|---|---|---|---|---|---|---|---|
| Biosynthesized IONP from the extract of Excoecaria cochinchinensis Lour leaves | Cd (II) | 8.07 | LGR | PSO | 18.32 | 2.5 g/L | - | (Lin, et al., 2018) |
| Extracts of Camellia sinensis (green tea, GT), Syzygium aromaticum (clove, CL), Mentha spicata (spearmint, SM), Punica granatum juice (pomegranate, PG), and Red Wine (RW), for the production of nano iron suspensions | Ni (II) | 6.0 | LGR | - | 227.20 | 0.05 g | Endothermic | (Nithya, et al., 2018) |
| IONP synthesized by eucalyptus leaf extract | Pb (II) | 2.0 | LGR | PSO | 18.69 | 1 mg/L | Endothermic and Spontaneous | (Cao, et al., 2016) |
| Green tea polyphenol NPs | Pb (II) | 5.5 | LGR | PSO | 584.8 | 20 mg/L | - | (Deng, et al., 2021) |
| Phytogenic zero-valent iron NPs reproducible Catharanthus roseus (CR) flower extract | Cr (VI) | 4.3 | LGR | PSO | 9.09 | 1.6 g/L | - | (Sravanthi, Ayodhya, and Swamy, 2019) |
| Magnetic IONP ($Fe_3O_4$) from tea waste | As (III) and As (V) | 7.0 | FRH | PSO | 188.69 and 153.8 | 1 g/L | Endothermic and spontaneous | (Lunge, Singh, and Sinha, 2014) |

| Green NMs | HMI | pH | IST | KNT | Maximum SNC (mg/g) | Dosage | TMD factors | References |
|---|---|---|---|---|---|---|---|---|
| Bioinspired 2D-carbon flakes and $Fe_3O_4$ NPs | As (III) | 7.0 | LGR | PSO | 57.47 | 0.1 g/L | Exothermic and spontaneous | (Venkateswarlu, Lee, and Yoon, 2016) |
| M. oleifera leaf extract capped magnetic NPs | Pb (II) | 5.0 | FRH | PSO | 49.00 | 0.2 g/L | Endothermic and spontaneous | (Gautam, Shivalkar, and Banerjee, 2020) |
| Encapsulated green magnetic NPs | Pb (II) and Cd (II) | 6.0 | LGR | PSO | 548.00 and 610.67 | 0.03 g/L | Exothermic and spontaneous 0.03 | (Ali, et al., 2019) |

## 6.3 Factors Impacting the Adsorption of Pollutants

Several factors can affect the course of sorption. The value of pH has an incredible impact on the sorption course as it influences the solution chemical state along with the sorbent chemical state and its surface (Gad and Awwad, 2007). From the perspective of cost, a system would be able to confiscate contaminants over a widespread pH (Ali, et al., 2017).

With the value of pH at a zero-point charge ($pH_{PZC}$) and sorbate charge, it is probable to determine the solution pH value suitable for a particular sorbate removal. If the value of the pH is lower than the $pH_{PZC,}$ the SS is +ve charged and attracts anions and when the pH is higher than the $pH_{PZC,}$ the SS is charged -ve and attracts cations. At a certain value of pH, precipitation may happen and the results from the sorption process may therefore be incorrect (Pfeifer and Škerget, 2020). Findings from papers reviewed in this book chapter show that the optimum elimination of most HMI was achieved at a low pH range of 2–6 and in other cases at pH 7–10 (Table 6.1).

Also, the sorption course may be transformed and impacted by altering the initial metal CCTN or the SD. These variables may decrease or increase sorption and in some instances, an elevated initial sorbate CCTN may depict a huge increase in the rate of sorption, as it creates an essential operating interaction to surmount the resistance of the liquid. The sorption rate could equally be reduced swiftly after initial sorption, owing to the dynamic locations that are simply accessible to be inhabited, triggering a minor alteration on the charge on the SS (Pfeifer and Škerget, 2020). It was noticed in these studies that the removal efficiency of the sorption process decreased with enhanced initial sorbate CCTN. This was attributed to the increased collision between the sorbate and sorbent as the sorbate's initial CCTN is increased, leading to an enhanced rate of aggregation and improved surface area for HMI sorption (Fazlzadeh, et al., 2017 and Sravanthi, Ayodhya, and Swamy, 2019).

Improvement in the amount of HMI removed as the dosage of the sorbent employed was increased was reported in the studies by (Sravanthi, Ayodhya, and Swamy, 2019; Fazlzadeh, et al., 2017; and Ifijen, et al., 2020). With an increase in the dosage of sorbent used, a great number of active sites for binding are produced, and the rate of sorption increases to a certain degree. Also, when the dosage of the sorbent employed is very high, the rate of sorption is affected by it due to particles aggregated created and the reduced number of the active site available for MI sorption (Pfeifer and Škerget, 2020). Also represented in Table 6.1 are the optimum dosages employed for the removal of various HMI.

## 6.4 Isotherm and Kinetics Models

To better demonstrate the sorption conduct of green NMs using some sorption IST and KNT models are briefly introduced in this book chapter. The evaluation of the sorbent sorption capacities and the establishment of the sorption equilibrium, where the correlation between the amount of the sorbate on the sorbent and the sorbates equilibrium CCTN under uniform temperatures is termed sorption IST. The various types of models of IST mostly employed in the sorption process are LGR, FRH, Dubinin-Radushkevich (DR), Sips, Temkin models, etc. The most all-embracing models employed are the FRH and the LGR models and are rigorously reviewed in this chapter (Yang, et al., 2019). The LGR model is utilized to calculate approximately the sorbent SNC and it is suggestive of the uptake that happens on an identical surface by monolayer sorption devoid of interaction between the sorbed molecules. The linearized procedure of the LGR model is described by Equation (6.1);

$$\frac{C_e}{q_e} = \frac{1}{bq_m} + \frac{C_e}{q_m} \tag{6.1}$$

$C_e$, $q_e$, $q_m$, and $b$ represent the equilibrium CCTN (mg/L), the quantity of sorbate sorbed per unit of the sorbent at equilibrium (mg/g), the maximum SNC of sorbate per unit of sorbent (mg/g), and the constant of LGR sorption (L/mg).
The important attributes of an LGR IST can be stated in terms of a non-dimension constant separation parameter ($R_L$) given by Equation (6.2).

$$R_L = \frac{1}{1 + bC_o} \tag{6.2}$$

$C_0$ represents the initial sorbate CCTN in mg/L. Sorption is favorable, unfavorable, linear, and irreversible when (Jin, et al., 2018).
While in the FRH model, multilayer sorption assumes a non-uniform distribution of sorption heat and similarities over a diverse SS. The linearized form of this model is given in Equation (6.3).

$$\log q_e = \log K_F + \frac{1}{n}\log C_e \tag{6.3}$$

$K_F$ and $\frac{1}{n}$ define the SNC and the heterogeneity factor whose value specifies if the sorption process is irreversible $\left(\frac{1}{n} = 0\right)$, favourable $\left(\frac{0<1}{n<1}\right)$, and unfavourable $\left(\frac{1}{n} > 1\right)$ (Abouzeid, et al., 2018).

The intent of KNT is important for the sorption systems and rate-controlling step design as the chemical reaction transpires. It also describes the interactions between the number of sorbates sorbed on the sorbent ($q_t$) and the time of contact ($t$). The most utilized sorption KNT models are the pseudo-first-order (PFO) and PSO models, with the PFO model relating to the sorption of liquid-solid phase systems centered on the SNC. Whereas, the PSO describes the

SNC of the solid stages. The linearized form of both KNT models is expressed by Equations (6.4) and (6.5).

$$\log(q_e - q_t) = \log q_e - \frac{k_1}{2.303}t \tag{6.4}$$

$$\frac{t}{q_t} = \frac{1}{K_2 q_e^2} + \frac{1}{q_e}t \tag{6.5}$$

where $q_e$, $q_t$, $K_1$, and $K_2$ represent the equilibrium SNC of the sorbent (mg/g), the SNC at a contact time *t* (mg/g), the rate constants of the PFO and PSO models (Yang, et al., 2019).

The sorption process employed to study the confiscation performance as well as the machineries in the research papers reviewed in this book chapter, due to the rate-controlling step in the sorption presents suggestions to recognize the mechanism of removal (Ali, et al., 2017). Hence, the various forms of sorption IST models employed and outcomes were compared utilizing linear regression correlation coefficient values ($R^2$). Reviewed papers show that the PSO and LGR models best portrayed the sorption KNT (which supported the assumption of sorption was chemisorption) and IST data (monolayer sorption of contaminants/sorbates on a solid surface) in most reports highlighted in Table 6.1.

## 6.5 Thermodynamics

The sorption TMD process is studied to determine the sorption process mechanism, whether endothermic or exothermic. The change in Gibbs free energy (ΔG°) and the Van't Hoff equation allow for the change in entropy (ΔS°) and change in enthalpy (ΔH°) determination, given by Equations (6.6) and (6.7).

$$\Delta G^{\circ\circ} = -RT \ln K \tag{6.6}$$

$$\Delta G^{\circ} = (\Delta H^{\circ} - T)\,\Delta S^{\circ} \tag{6.7}$$

Putting Equation (6.6) into Equation (6.7), Equation (6.8) is obtained, which is,

$$-RT \ln K = (\Delta H^{\circ} - T)\,\Delta S^{\circ} \tag{6.8}$$

$$\ln K = \frac{\Delta S^{\circ}}{R} - \frac{\Delta H^{\circ}}{RT} \tag{6.9}$$

A linear plot of In K against T gives the values of $\Delta H^\circ$ and $\Delta S^\circ$ from the slope and intercept of the fitted curve. R, T, and K represent the universal gas constant (8.314 J/mol K), the temperature in Kelvin, and the TMD equilibrium constant (Lyubchik, et al., 2011). The value of $\Delta G^\circ$ can distinguish whether a procedure is spontaneous or not and its -ve value indicates a spontaneous process. While $\Delta H^\circ$ give evidence on the nature of the sorption process whether exothermic or endothermic and it is used to distinguish between physical and chemical sorption process. Also, the extent of changes in the SS is predicted by $\Delta S^\circ$, owing to the changes observed in the reversibility, which is impacted by the -ve value of $\Delta S^\circ$ acquired (Rincón-Silva, Moreno-Piraján, and Giraldo, 2015). The sorption of As (V) to GNMs was detected to be spontaneous, exothermic, and random at the boundary of the solid/solution during sorption (Mukherjee, et al., 2016). Overall, as observed from papers reviewed in this book chapter, most HMI sorption processes were spontaneous and both exothermic and endothermic in nature (see Table 6.1).

## 6.6 Conclusions and Future Outlook

NT has been well-thought-out and efficient in solving issues relating to water quality and quantity. There have been significant variations in cutting-edge NT that nurture new options for the development of improved water and wastewater technology processes. In that regard, this chapter highlights the different green NMs synthesized via the green approach and their use as sorbents for the decontamination of wastewater contaminated with HMI owing to their simple, non-noxiousness, cost-effective and eco-friendly approach for the synthesis of NPs. Future studies should focus on improving the commercial feasibility of green NMs and assessing water treatment systems' collaborative processes. Also, their probable noxiousness to the well-being of humans and the ecosystem should be fully assessed seriously to safeguard their applications

It is therefore apparent that GNMs vis-à-vis GNT are innovative means for environmental sustainability that need to be appropriately explored. Consequently, the future stance should consider and embolden the integration of GNMs for HMI removal as a necessity for environmental management considering its numerous advantages compared to other means as this will also be of great significance in guaranteeing effective monitoring, treatment, and usage of wastewater and HMI.

## Acknowledgment

The authors are grateful to their respective institutions and authors whose publications were used for writing this chapter.

## References

1. Abdel-Aziz, H.M., R.S. Farag, and S.A. Abdel-Gawad. 2019. "Carbamazepine removal from aqueous solution by green synthesis zero-valent iron/Cu nanoparticles with Ficus Benjamina leaves' extract." *International Journal of Environmental Research* **13** (5): 843–852.
2. Abouzeid, R.E., R. Khiari, N. El-Wakil, and A. Dufresne. 2018. "Current state and new trends in the use of cellulose nanomaterials for wastewater treatment." *Biomacromolecules* **20** (2): 573–597.
3. Aigbe, U.O., and O.A. Osibote. 2020. "A review of hexavalent chromium removal from aqueous solutions by sorption technique using nanomaterials." *Journal of Environmental Chemical Engineering* **8** (6): 104503.
4. Aigbe, U.O., M.K., Ho W.H. Khenfouch, A. Maity, V.J. Vallabhapurapu, and N.M. Hemmaragala. 2018. "Congo red dye removal under the influence of rotating magnetic field by polypyrrole magnetic nanocomposite." *Desalination and Water Treatment* 131: 328–342.
5. Aigbe, U.O., R. Das, W.H. Ho, V. Srinivasu, and A. Maity. 2018. "A novel method for removal of Cr (VI) using polypyrrole magnetic nanocomposite in the presence of unsteady magnetic fields." *Separation and Purification Technology* 194: 377–387.
6. Aigbe, U.O., R.B. Onyancha, K.E. Ukhurebor, and K.O. Obodo. 2019. " Removal of fluoride ions using a polypyrrole magnetic nanocomposite influenced by a rotating magnetic field." *RSC Advances* **10** (1): 595–609.
7. Aigbe, U.O., W.H. Ho, A. Maity, M. Khenfouch, and V. Srinivasu. 2018. "Removal of hexavalent chromium from wastewater using PPy/$Fe_3O_4$ magnetic nanocomposite influenced by rotating magnetic field from two pole three-phase induction motor." *Journal of Physics: Conference Series* **984** (1): 012008.

8. Ali, I, C. Peng, D. Lin, D.P. Saroj, I. Naz, Z.M. Khan, M. Sultan, and M. Ali. 2019. "Encapsulated green magnetic nanoparticles for the removal of toxic $Pb^{2+}$ and $Cd^{2+}$ from water: development, characterization and application." *Journal of Environmental Management* 234: 273–289.

9. Ali, I., C. Peng, I. Naz, Z.M. Khan, M. Sultan, T. Islam, and I.A. Abbasi. 2017. "Phytogenic magnetic nanoparticles for wastewater treatment: A review." *RSC Advances* **7** (64): 40158–40178.

10. Al-Qahtani, K.M. 2017. "Cadmium removal from aqueous solution by green synthesis zero valent silver nanoparticles with Benjamina leaves extract." *The Egyptian Journal of Aquatic Research* **43** (4): 269–274.

11. Bahrulolum, H, S. Nooraei, N. Javanshir, H. Tarrahimofrad, V.S. Mirbagheri, A.J. Easton, and G. Ahmadian. 2021. "Green synthesis of metal nanoparticles using microorganisms and their application in the agrifood sector." *Journal of Nanobiotechnology* **19** (1): 86.

12. Barman, S.R., U. Roy, P. Das, and A. Mukhopadhayay. 2021. "Membrane processes for removal of polyaromatic hydrocarbons from wastewater." *Green Chemistry and Water Remediation: Research and Applications* 189–207.

13. Cao, D., X. Jin, L. Gan, T. Wang, and Z. Chen. 2016. "Removal of phosphate using iron oxide nanoparticles synthesized by eucalyptus leaf extract in the presence of CTAB surfactant." *Chemosphere* 159: 23–31.

14. Das, C., S. Sen, T. Singh, T. Ghosh, S.S. Paul, T.W. Kim, S. Jeon, D.K. Maiti, J Im, and G Biswas. 2020. "Green synthesis, characterization and application of natural product coated magnetite nanoparticles for wastewater treatment." *Nanomaterials* **10** (8): 1615.

15. Deng, Z., Z. Yi, G. Chen, X. Ma, Y. Tang, and X. Li. 2021. "Green tea polyphenol nanoparticle as a novel adsorbent to remove $Pb^{2+}$ from wastewater." *Materials Letters* 284: 128986.

16. Ehrampoush, M.H., M. Miria, M.H. Salmani, and A.H. Mahvi. 2015. "Cadmium removal from aqueous solution by green synthesis iron oxide nanoparticles with tangerine peel extract." *Journal of Environmental Health Science and Engineering* **13** (1): 1–7.

17. Fazlzadeh, M., K. Rahmani, A. Zarei, H. Abdoallahzadeh, F. Nasiri, and R. Khosravi. 2017. "A novel green synthesis of zero valent iron nanoparticles (NZVI) using three plant extracts and their efficient application for removal of Cr (VI) from aqueous solutions." *Advanced Powder Technology* **28** (1): 122–130.

18. Gad, H.M.H., and N.S. Awwad. 2007. "Factors affecting on the sorption/desorption of Eu (III) using activated carbon." *Separation Science and Technology* **42** (16): 3657–3680.

19. Gangadhar, G., U. Maheshwari, and S. Gupta. 2012. "Application of nanomaterials for the removal of pollutants from effluent streams." *Nanoscience and Nanotechnology-Asia* **2** (2): 140–150.

20. Gautam, P.K., S. Shivalkar, and S. Banerjee. 2020. "Synthesis of M. oleifera leaf extract capped magnetic nanoparticles for effective lead [Pb (II)] removal from solution: Kinetics, isotherm and reusability study." *Journal of Molecular Liquids* 305: 112811.

21. Hashem, E.A. 2014. "Nanotechnology in water treatment, case study." *Egyptian Journal of Economic Development Studies* **2** (3): 243–259.

22. Ifijen, I.H., A.B. Itua, M. Maliki, C.O. Ize-Iyamu, S.O. Omorogbe, A.I. Aigbodion, and E.U. Ikhuoria. 2020. "The removal of nickel and lead ions from aqueous solutions using green synthesized silica microparticles." *Heliyon* **6** (9): e04907.

23. Jawed, A., V. Saxena, and L.M. Pandey. 2020. "Engineered nanomaterials and their surface functionalization for the removal of heavy metals: A review." *Journal of Water Process Engineering* 33: 101009.

24. Jin, X., Y. Liu, J. Tan, G. Owens, and Z. Chen. 2018. "Removal of Cr (VI) from aqueous solutions via reduction and absorption by green synthesized iron nanoparticles." *Journal of Cleaner Production* 176: 929–936.

25. Karman, S.B., S.Z.M. Diah, and I.C. Gebeshuber. 2015. "Raw materials synthesis from heavy metal industry effluents with bioremediation and phytomining: A biomimetic resource management approach." *Advances in Materials Science and Engineering* **2015** (3): 1–21

26. Lin, J., B. Su, M. Sun, B. Chen, and Z. Chen. 2018. "Biosynthesized iron oxide nanoparticles used for optimized removal of cadmium with response surface methodology." *Science of the Total Environment* 627: 314–321.

27. Lingamdinne, L.P., J.R. Koduru, and R. Rao Karri. 2019. "Green synthesis of iron oxide nanoparticles for lead removal from aqueous solutions." In *Key Engineering Materials*, 122–127. Trans Tech Publications Ltd.

28. Lunge, S., S. Singh, and A. Sinha. 2014. "Magnetic iron oxide ($Fe_3O_4$) nanoparticles from tea waste for arsenic removal." *Journal of Magnetism and Magnetic Materials* 356: 21–31.

29. Lyubchik, S., A. Lyubchik, O. Lygina, S. Lyubchik, and I. Fonseca. 2011. "Comparison of the thermodynamic parameters estimation for the adsorption process of the metals from liquid phase on activated carbons." In *Thermodynamics-Interaction Studies-Solids, Liquids and Gases*. IntechOpen.

30. Mahanty, S., S. Chatterjee, S. Ghosh, P. Tudu, T. Gaine, M. Bakshi, S. Das, et al. 2020. "Synergistic approach towards the sustainable management of heavy metals in wastewater using mycosynthesized iron oxide nanoparticles: Biofabrication, adsorptive dynamics and chemometric modeling study." *Journal of Water Process Engineering* 37: 101426.

31. Mukherjee, D., S. Ghosh, S. Majumdar, and K. Annapurna. 2016. "Green synthesis of α-$Fe_2O_3$ nanoparticles for arsenic (V) remediation with a novel aspect for sludge management." *Journal of Environmental Chemical Engineering* **4** (1): 639–650.

32. Nasrollahzadeh, M., M. Sajjadi, S. Iravani, and R.S. Varma. 2021. "Green-synthesized nanocatalysts and nanomaterials for water treatment: Current challenges and future perspectives." *Journal of Hazardous Materials* 401: 123401.

33. Nazal, K. 2019. *Marine algae bioadsorbents for adsorptive removal of heavy metals.* IntechOpen.

34. Nithya, K., A. Sathish, P.S. Kumar, and T. Ramachandran. 2018. "Fast kinetics and high adsorption capacity of green extract capped superparamagnetic iron oxide nanoparticles for the adsorption of Ni (II) ions." *Journal of Industrial and Engineering Chemistry* 59: 230–241.

35. Onyancha, R.B., U.O. Aigbe, K.E. Ukhurebor, and P.W. Muchiri. 2021. "Facile synthesis and applications of carbon nanotubes in heavy-metal remediation and biomedical fields: A comprehensive review." *Journal of Molecular Structure* 1238: 130462.

36. Pfeifer, A., and M. Škerget. 2020. "A review: A comparison of different adsorbents for removal of Cr (VI), Cd (II) and Ni (II)." *Turkish Journal of Chemistry* **44** (4): 859–883.

37. Puthukkara, A.R.P., S.T. Jose, and D.S. Lal. 2021. "Plant mediated synthesis of zero valent iron nanoparticles and its application in water treatment." *Journal of Environmental Chemical Engineering* **9** (1): 104569.

38. Rincón-Silva, N.G., J.C. Moreno-Piraján, and L.G. Giraldo. 2015. "Thermodynamic study of adsorption of phenol, 4-chlorophenol, and 4-nitrophenol on activated carbon obtained from eucalyptus seed." *Journal of Chemistry* **2015** (6): 1–12

39. Sharma, G., B. Thakur, M. Naushad, H. Ala'a, A. Kumar, M. Sillanpaa, and G.T. Mola. 2017. "Fabrication and characterization of sodium dodecyl sulphate@ ironsilicophosphate nanocomposite: Ion exchange properties and selectivity for binary metal ions." *Materials Chemistry and Physics* 193: 129–139.

40. Singh, J., T. Dutta, K.H. Kim, M. Rawat, P. Samddar, and P. Kumar. 2018. "'Green'synthesis of metals and their oxide nanoparticles: Applications for environmental remediation." *Journal of Nanobiotechnology* **16** (1): 1–24.

41. Singh, P., S.K. Yadav, and M. Kuddus. 2020. "Green nanomaterials for wastewater treatment." In *Green Nanomaterials* , 227–242. Singapore: Springer.

42. Sravanthi, K., D. Ayodhya, and P.Y. Swamy. 2019. "Eco-friendly synthesis and characterization of phytogenic zero-valent iron nanoparticles for efficient removal of Cr (VI) from contaminated water." *Emergent Materials* **2** (3): 327–335.

43. Ukhurebor, K.E., U.O. Aigbe, R.B. Onyancha, W. Nwankwo, O.A. Osibote, H.K Paumo, O.M Ama, C.O Adetunji, and I.U Siloko. 2021. "Effect of hexavalent chromium on the environment and removal techniques: A review." *Journal of Environmental Management* 280: 111809.

44. Venkateswarlu, S., D. Lee, and M. Yoon. 2016. "Bioinspired 2D-carbon flakes and $Fe_3O_4$ nanoparticles composite for arsenite removal." *ACS Applied Materials and Interfaces* **8** (36): 23876–23885.

45. Wan, H., L. Nan, H. Geng, W. Zhang, and H. Shi. 2021. "Green synthesis of a novel MXene–CS composite applied in treatment of Cr (VI) contaminated aqueous solution." *Processes* **9** (3): 524.

46. Weng, X., Z. Lin, X. Xiao, C. Li, and Z. Chen. 2018. "One-step biosynthesis of hybrid reduced graphene oxide/iron-based nanoparticles by eucalyptus extract and its removal of dye." *Journal of Cleaner Production* 203: 22–29.

47. Wenten, I.G. 2016. "Reverse osmosis applications: prospect and challenges." *Desalination* 391: 112–125.

48. Xiao, X., Q. Wang, G. Owens, F. Chiellini, and Z. Chen. 2019. "Reduced graphene oxide/iron nanoparticles used for the removal of Pb (II) by one step green synthesis." *Journal of Colloid and Interface Science* 557: 598–607.

49. Yang, J., B. Hou, J. Wang, B. Tian, J. Bi, N. Wang, X. Li, and X. Huang. 2019. "Nanomaterials for the removal of heavy metals from wastewater." *Nanomaterials* **9** (3): 424.

## Chapter 7

# Significant Role of Green Nanomaterials in Wood-Based Industries: Environmental and Quality Strategies

**Ranjana Yadav**
*Department of Chemistry, Indian Plywood Industries Research and Training Institute, Centre Mohali, Punjab, India*
ranjanay7@gmail.com

A vast range of sophisticated materials has radically altered the landscape of science and technology during the last few decades. In most cases, nanotechnology is used as a catch-all term for all kinds of nanoscale material research. Because of their independence from petrochemical origin, biobased green materials give up a whole new world of possibilities in each industry. Researchers have investigated the use of environmentally friendly materials for wood glues as a result of increased concerns about natural imprints and the harmfulness of synthetic adhesives. In comparison to formaldehyde-based wood adhesives, nanoparticles (NPs) made from plant extracts, nanocellulose (NC), and lignin NPs (LNPs) are a more sustainable and cost-effective bionanomaterial. Green NPs can be used as fasteners and fortification in several synthetic resin processes to improve quality. The wood-based products industry

*Green Nanoarchitectonics*
Edited by Kaushik Pal

ISBN 978-981-4968-50-8 (Hardcover), 978-1-003-31860-6 (eBook)
www.jennystanford.com

might also make use of a variety of widely available nanomaterials to improve the performance of existing products or develop new forest-based value-added products. This review chapter mainly focused on the application of renewable resourced nanomaterials and executes their new role in wood-based industry products.

## 7.1 Introduction

In wood adhesive innovation, synthetic resins are the overwhelming adhesive for the development of different wood boards [1]. Formaldehyde emissions are primarily due to unreacted free formalin and moderate adhesive hydrolysis under hot/humid conditions during the manufacturing and utilization of boards [2, 3]. Due to formalin outflows, it has been demonstrated that formaldehyde-based adhesives are not harmless to ecosystem items [4]. Subsequently, to decrease free formalin and volatile organic substance discharge and for sustainable development, biobased materials could be one of the choices past their inexhaustibility [5–7].

Research has attracted extraordinary attention in generally biobased adhesives, for example, starch [8–12], soy protein [13, 14], or inexhaustible elastic, different cellulosic materials [15, 16], for application in the glue area. Among these biopolymers, cellulose is the most bountiful sustainable biomaterial, and its normal preference for self-adhesion makes it a latent material in adhesion discipline. They might be cellulose nanocrystals (NCCs or CNCs), nanofibrillated celluloses (CNF) and bacterial nanocellulose (BC) [17]. It will in general be accomplished through mechanical or chemical treatments [18]. NC has uncommon characteristics compared with bulk materials [19], and the bonding strength of NC has been explored in different examinations [20, 21].

Nanotechnology is in effect intensively produced for a wide assortment of utilizations, from material behavior improvement to medical uses [22]. Nanotechnology is a quickly developing study of delivering and using nano measures particles. The field of nanotechnology is the most unique area of research in material sciences, and the preparation of nanoparticles (NPs) is getting fundamentally all through the world [23–26]. There is a wide scope of utilization for such environmentally biodegradable nanocomposites synthesized from renewable natural nanomaterials, for example, NC and lignin NPs (LNPs) [27–29].

Silver NPs have further developed wood protection [30–32], conduct against fire [33], physical properties [33, 34], thermal treatment [35], densification [36], particleboard (PB) production [37], and drying [25]. There are various techniques for synthetizing silver NPs [38]. Biological preparation of nanoparticles includes utilization of biomaterials, for example, bacteria [39], fungi, yeast, viruses, microalgae [40, 41], and plant extracts [42]. Of every biomaterial, plant extracts are viewed as more favorable for the preparation because of their accessibility, sustainable behavior, effortlessness of cycle, productivity, frequency and stability of incorporated NPs and rate viability [43–48].

This review chapter mainly focuses on the synthesis of green nanomaterials and their role in wood-based industrial products.

## 7.2 Resource of Green Nanoparticles

Nanoparticles can be utilized as fillers or adhesives in different polymers, so various upgrades in material properties can be accomplished. They can likewise be utilized to build up thermosetting polymers to further develop the last execution properties. Hence, NPs are receiving expanded interest for innovative work. The impact of NP-altered thermosetting adhesives on the mechanical properties of wood composite boards has been assessed by a few writers [14, 49–51].

The two techniques for developing NPs are "top-down" and "bottom-up" [52, 53]. In the top-down methodology, the appropriate bulk material is reduced in size using various strategies, such as pulse laser removal, evaporation–condensation, ball processing, pulse wire release strategy and so on, in the bottom-up approach. NPs can be incorporated utilizing chemical and biological strategies without anyone else getting together the marvel of iotas to new cores that develop into a molecule of nanoscale. In top-bottom methodology, evaporation-condensation is the broadest strategy for the production of nanoparticles.

The traditional techniques for the synthesis of NPs are costly, poisonous, and nonclimate cordial. To address these issues, specialists have identified the precise green courses or naturally occurring sources and their products that can be utilized to produce nanoparticles. Green synthesis can be classified as (a) the use of microorganisms such as fungi, yeasts, microscopic organisms, and actinomycetes, (b) plants and plant extricates, and (c) the utilization of layouts such as films, viruses, and diatoms. Green synthesis through microscopic organisms, fungi, plants, and plant extracts is depicted in further sections (Fig.7.1).

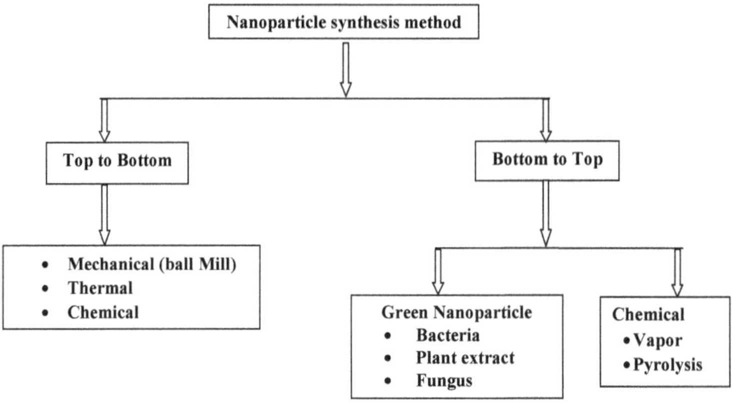


**Figure 7.1** Different method of nanoparticle synthesis.

**Table 7.1** List of plants used for the production of silver NPs

| Species | Plant's part used for synthesis |
|---|---|
| *Ipomoea Pes-Caprae* | Stem |
| *Olea ferruginea* | Leaf and bark |
| *Phyllanthus emblica* | Fruit |
| *Salvia spinosa* | Plant |
| *Piper pedicellatum* | Leaves |
| *Moringa aleifera* | Flowers |
| *Artocarpus heterophyllus* | Seeds |
| *Aloe vera* | Leaves |
| *Emblica officinalis* | Fruit |
| *Leucas aspera* | Bark |
| *Murraya koenigii* | Leaves |
| *Terminalia chebula* | Fruit |
| *Tectona grandis* | Wood flour |
| *Moringa oleifera* | Leaf |

| Species | Plant's part used for synthesis |
|---|---|
| *Cucumis prophetarum* | Leaf |
| *Viola serpens* | Leaves |
| *Ocinum tenuiflorum* | Leaves |

*Source*: Refs. 54–57.

**Table 7.2** List of plants used for ZnO NPs

| Plant species | Plant's part used for synthesis |
|---|---|
| *Citrus reticulate* | Fruit peel |
| *Justicia adhatoda* | Leaves |
| *Xanthomonas oryzae* | Plant |
| *Lycopersicon esculentum M.* | Tomato leaf |
| *Rosa canina* | Leaves |
| *Moringa oleifera* | Leaves |
| *Nephelium lappaceum* | Peel |
| *Plectranthus amboinicus* | Leaves |
| *Solanum nigrum* | Leaves |
| *Cocus nucifera* | Coconut water |
| *Vitex negundo* | Flowers |
| *Trifolium pratense* | Flowers |
| *Passiflora caerula* | Leaves |
| *Carica papaya* | Leaves |
| *Calotropis procera* | Leaves |
| *Deverra tortuosa* | Aerial parts |
| *Myristica fragrans* | Fruit |
| *Laurus nobilis L.* | Leaves |
| *Phyllanthus nirur* | Leaves |
| *Rosa canina* | Fruit |
| *Mussaenda frondosa L.* | Leaf and stem |
| *Cayratia pedata* | Leaf |
| *Azadirachta indica* | Leaves |
| *Aloe Vera* | Leaves |
| *Matricaria chamomilla L.* | Flower |
| *Olea europaea* | Leave |

*Source*: Refs. 56, 58, 59, and 60.

### 7.2.1 Nanoparticles by Using Plant and Plant Extracts

Copper oxide NPs were prepared utilizing leaf concentrates from chosen plants and orange peel extract. A solution of $CuSO_4.5H_2O$ (0.025 M) was preheated separately in a hot water bath at 60°C for 30 minutes before being combined at 1000 rpm for 10 minutes at a ratio of 1:4 [61]. A change in color from a somewhat blue green to a dull indicated the presence of copper oxide.

Phenols, carbonyl groups, flavonoids, alkaloids, terpenoids, proteins, shades, and other decreased specialists/stabilizers found in plant extracts stimulate Ag-NP biosynthesis [62–66]. In addition to Ag-NP synthesis, experts are looking into a possible green ZnO-Nos production process. In green synthesis, synthetic "impurities" utilized as reducing agents or stabilizers are subbed by various natural substances, such as concentrates, chemicals or proteins obtained from different plant species and plant parts, for example, leaves [67], roots [68], seeds [69], bark/ring [70], natural products [71], and 12 blossoms [72], similar to microorganisms, green growth, and organisms [73].

Zinc oxide NPs were created using an environmentally friendly concentrate of *D. tortuosa* aerial parts as a decreasing and covering specialist. The production of zinc oxide NPs (ZnO NPs) was verified utilizing different analyses, and subjective and quantitative examinations of plant exudates were performed [74].

The improvement of green innovation is producing the interest of analysts toward an eco-friendly and minimal expense strategy for the bio synthesis of NPs. It was seen that the *Azardiachta indica* leaf concentrate can reduce copper particles into copper nanoparticles within 5 minutes of response time and brown colored precipitate settling down. The biosynthesized NPs of copper oxide are described through different analyses. The molecular sizes of the CuO NPs were found to be 49 nm, 87 nm, 120 nm, and 324 nm.

Green synthesis utilizing plant and plant extracts gives off the impression of being quicker than different microorganisms, such as microscopic organisms and parasites. The utilization of plant and plant separates in green blends has drawn consideration as a result of its quick development, giving a single-step strategy, conservative convention, nonpathogenic, and ecofriendly for NP synthesis [44, 75, 76]. Schematic diagram of nanoparticle synthesis by plant/plant parts has been shown in Fig 7.2.

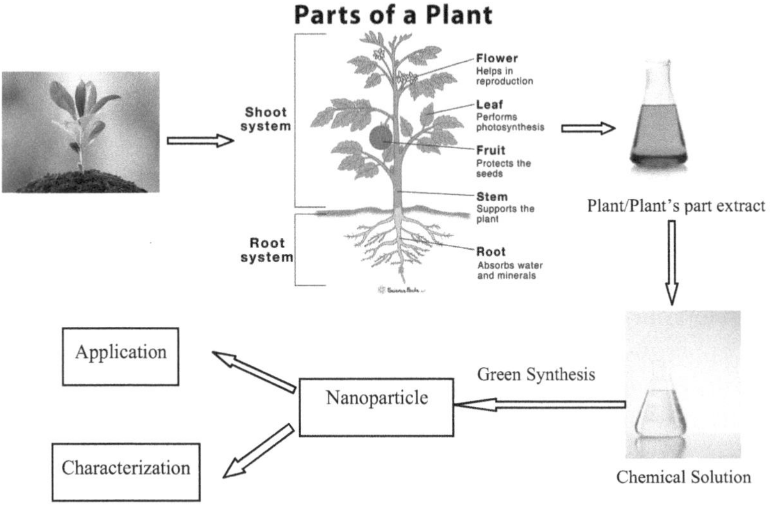


**Figure 7.2** Schematic diagram of nanoparticle synthesis by plant/plant parts.

## 7.2.2 Cellulose Nanoparticles

Nanocellulose is one of nature's predominant biodegradables and cost-effective nanomaterials. In more straightforward terms, NC compares to cellulose on the nanometer scale. People made cellulose from plant materials such as wood, plants, waste products, and algae until 1970 [77, 78]. To distinguish transparent NC fibers, cellulose must now undergo a combination of synthetic and mechanical treatments [78].

The advantages of utilizing NC as fortifications in resin for the creation of reconstructed wood boards include the chance of changing cement characteristics, acquiring the mechanical and actual properties of boards and decreasing formaldehyde discharges by utilizing petroleum-based resin [79]. The reinforcing character of cellulose nanofibers (CNFs) was demonstrated by their expansion in amino resin [80]. NC can be made from plant cellulose or as a result of essential metabolic cycles of certain bacterial strains [81]. BC has been broadly utilized in a variety of areas, from the food and paper industry to biomedical applications, because of its hydrophilic nature, flexibility, nonpoisonousness, great biocompatibility, and wide accessibility [82].

### 7.2.3 Lignin Nanoparticles

LNPs were synthesized after an adjusted convention of treating introductory lignin solutions in DMSO with water as an anti-solvent to dissolvable utilizing dialysis tubes made of recovered cellulose, as depicted [83–85]. Moreover, LNP divisions from beech and chestnut AL were prepared for successive dip costing.

Lignin nanoparticles can be prepared from specialized lignin obtained from the paper industry or second-generation ethanol biorefineries, and they have been proposed for a variety of applications, such as dynamic segments or fillers in polymer blends, displaying lignin properties such antioxidant action or UV protection on a nanoscale level [86].

Lignin was extracted from saw dust from beech and chestnut trees in the neighborhood using moderate acidolysis, which is a high-volume by product of the wood industry. The Klason lignin (KL) content and polysaccharide composition were resolved utilizing traditional strategies. Lignin nanoparticles were synthesized through a nondissolvable technique including dialysis and described by different spectroscopies. The larger part of the LNPs displayed a normal molecular distance of 90–120 nm [84].

## 7.3 Role of Nanomaterials in Wood-Based Industry Products

The popularity of wood in several industries, such as furniture assembly, development, and construction, prompted a demand for wood cement. The reinforcing of adhesive with NC has been identified as one of the few opportunities afforded by nanotechnology for forest wood and has been researched [87]. The physical and mechanical characteristics of the boards have improved as a result [88].

Wood-based items are useful; for example, the utilization of nanomaterials in wood coatings, for example, nano $TiO_{2,}$ will upgrade the usefulness of wood as far as its strength, fire retardancy and ultraviolet absorption diminish the water resistance [89–91]. The use of nanomaterials in wood preservation [30, 92] will work on the impregnation of wood with pesticides by guaranteeing that synthetic compounds can enter further into the wood, in this manner decreasing the issue of excessive leaching [93]. Researchers have published many review papers on the utilization of nano techniques in wood and wood-based products [94, 95].

This review chapter will be based on the new advancements in nanotechnology applications in the wood-based industry, mainly for wood composites, wood coatings, and wood strength (Fig.7.3).

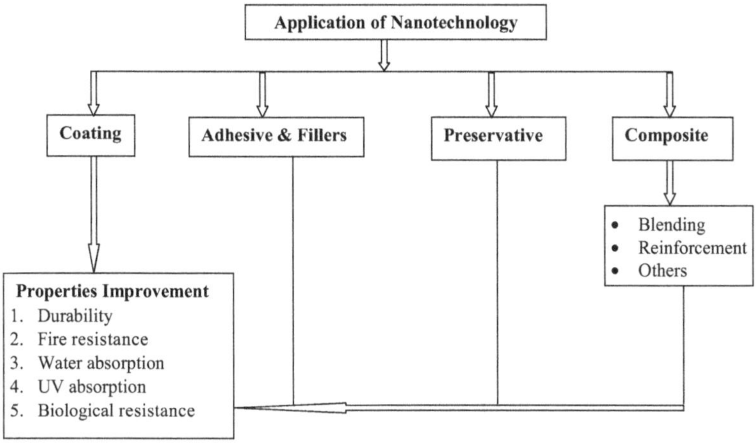


**Figure 7.3** Application of nanotechnology in wood-based industry.

## 7.3.1 Nanoparticles for Property Enhancement

Wood composites are by and large depicted a huge extent of items having a blend of wood components held together by a blinder. Among the benefits of wood composites are that they can be intended for explicit characteristics or execution prerequisites at various thicknesses, grades and sizes.

Wood composites are fabricated to take advantage of the natural strength characteristics of wood. The glues utilized released harmful free formalin in the final product. Nanotechnology can be used to work on the nature of wood-based composites to satisfy the expanding interest for existing items and for new items to be utilized in new applications.

Zinc oxide nanoparticles show great antibacterial activity. These NPs are mixed into melamine-urea formaldehyde (MUF) adhesive before being used to make PB [96, 97]. The findings demonstrate that the PB improved resistance to Gram-positive bacterium

*Staphylococcus aureus*, the Gram-negative bacterium *Escherichia coli*, the molds *Aspergillus niger* and *Penicillium brevicompactum* and the brown-decay parasite *Coniophora puteana*. When applied to the melamine-laminated surfaces of PBs, silver NPs, a well-known biocide adhesive, showed comparable antibacterial and anti-mold efficacy [98].

The combination of nano copper oxide and alkane surfactant was likewise affirmed to further develop water and termite resistance in treated compressed wood pieces [99]. PB fortified with changed polyvinyl acetate and oil palm starch, nano silicon oxide ($SiO_2$) and boric acid was found to have more decay resistance than wood composites reinforced with their natural starch [100].

Wood composite boards can be made by creating techniques to reduce the resin's cure time during hot pressing, thereby speeding up production and improving board quality. The expansion of ZnO NPs expanded the heat transfer at the PB's center point, generating a high level of resin cure and improving the physicomechanical characteristics [101].

Additionally, high conductivity NPs such as multiwalled carbon nanotubes (CNTS) and $Al_2O_3$ have been shown to improve the thermal and mechanical properties of MDF boards [102].
Several studies have focused on nanomaterial-supported wood composites that have enhanced physical and mechanical properties while emitting less formaldehyde. The use of nanoparticles to modify UF adhesive for plywood made from three wood species from quick development estates [103]. The modulus of rupture and elasticity of the board were improved by nanomodification of the resin with nanoclay at 0.75% [104].

Surprisingly, transition metal ion-modified bentonite (TMIBNT) nanoclay was utilized to convert crystalline UF resins to nebulous polymers using an in situ intercalation approach to prevent hydrogen bonding [105]. This brought about a 56.4% expansion in the bond strength and a 48.3% decrease in the formalin emission.

More Ag-NP formulations have been tested as a replacement for traditional wood preservatives. One of the initial examinations concerning the possible utilization of silver as biocide for wood assurance was made [106], which noticed that high proficiency of colloidal and ionic silver infused into wood structure had high efficacy bacteria.

Ag-NPs at concentrations of 200 and 400 ppm were used to protect poplar wood (*Populous deltoides*) from the white-rot fungus *Trametes versicolor*. Although the silver NPs could not completely inhibit fungal development, all things considered the weight misfortunes (WL) of treated wood dropped from 41.8% to 36.1% or to 29% at decay [107].

The decay activity of the white-decay organism *T. versicolorin* was completely disrupted by Ag-NPs administered at a concentration of 50 ppm by pressure impregnation. The weight loss of altogether tropical woods *Acacia magnium*, *Cedrela odorata*, and *Vochysia guatemalensis* decreased from 25–50% to under 5% for all of the species [108] at 4 months. The results demonstrated that the strength and resistance of these individual tropical wood species to the white-rot fungus had improved [109].

Ag-NPs against brown-rot fungi likewise archived [110] when their 1000 ppm and 3000 ppm concentrations did not repress but only retarded the growth of the fungus *Poria placenta*, with weight loss of wood samples decreasing by 13% or 7%, respectively. Additionally, wood-based composites, for example, PBs, can be saved from the activity of wood-decaying fungi [111, 112].

A viable antibacterial agent and ultraviolet absorber, ZnO NPs, are expected to be effective wood protectors against a variety of deteriorated wood chemicals. When nano zinc oxide (nano-ZnO) was used to modify lime wood either alone or in combination with ethylmethacrylate–methylacrylate copolymer, it had an antifungal effect of nano-ZnO against the wood-decaying fungi *Trametes versicolor* and *C. puteana* [113].

Previously reported data in decaying fungi noticed [114] by utilizing FTIR spectroscopy, a slight decrease in weight loss of approximately 9% for samples of *Pinus nigra* impregnated with ZnO nanoparticles, as indicated by the full-cell measurement, was seen by [115, 116] and established that examples of *Populus alba* treated with 1% and 2% ZnO-treated polyvinyl-butyral have good resistance against the white-rot fungus *T. versicolor*. Moreover, wood impregnation with ZnO NPs has no adverse effect on the bending strength to grains [115].

Developed the decay resistance of WPC against the degrading fungi *T. versicolor* and *C. puteana* when their weight loss was insignificant, close to 0% at 3 wt% expansion of ZnO NPs [117].

Similarly, when the immersion styrene solution was adjusted with ZnO NPs, a measurable improvement in WPC resistance to wood-rotting fungi was noticed [118]--weight loss of samples treated with 0.5 and 1.5 wt% content of ZnO NPs was brought down approximately 15% and 22% at activity the white-decay fungus *T. versicolor* and approximately 8–14% at activity the brown-rot fungus *C. puteana*, respectively. Degradation of PBs by wood-degrading fungi was clearly smothered, and UF resin was adjusted with 15 wt% of ZnO NPs [119] and correspondingly utilized MUF resin altered with 6–24 wt% ZnO NPs [120].

Zinc oxide NPs can be added to impregnating melamine formaldehyde resin at a content of 0.1–3 wt% utilized for surface lamination of PBs or other hardwood composites, with the goal of creating antimicrobial or antimicrobial-active surfaces [26].

To enhance the physical and mechanical characteristics of wood panels, nano wollastonite (NW) [121, 122], nano-ZnO [123], nano-$SiO_2$, nano-$Al_2O_3$ [124], nano silver [125], and NC (NCC and NFC) were incorporated. Nano cellulose was used as a filler and additive for the resin. In view of the examinations, the expansion NCC fundamentally worked on the mechanical characteristics of ply in the measure of 10% per 100 g of strong adhesive [126]. The qualities of wood particle composites with the addition of NCC to UF glues showed smooth surfaces and irrelevant distinctions in the density and moisture content of the boards [127, 128]. Indeed, even the screw withdrawal strength was shown to increase when the NFC expansion proportion and board thickness increased [129, 130].

Three diverse nanoscale consolidates have been tried: LNPs obtained structure beech wood through nonsoluble techniques, including dialysis; BC obtained from cultures fed with agro-nutritious waste; and CNCs synthesized from natural cellulose. The outcomes obtained for the three combines revealed significant contrasts: LNPs and CNCs penetrated the treated wood approximately a millimeter deep, while BC framed a conservative layer on the outside of the cell dividers all through the thickness of the examples [131].

Developed environmentally friendly and sustainable coatings of NC using spin- and dip-coating methods on two distinctive permeable cellulose substrates. Microscopy examinations showed that the spin-coating procedure was appropriate for substrates with more modest pore sizes, while dip-coating was reasonable for substrates with larger pore sizes. These outcomes demonstrate that NC coatings have extraordinary potential as feasible elective coatings on paperboard [132].

Because of the better nanoclay dispersion, which resulted in greater interfacial association with the resin, the nanoclay improved the mechanical properties of novolac-based phenolic resin (PF) and woven glass fiber (GF). The nanoclay increased the flexible modulus by 38% for CN and 43% for CB at a concentration of 2.5 wt%. PF and glass fiber composites were prepared for aging procedures with several aqueous solutions, such as water, saline solution, and acidic conditions, and the upgrades further developed solidness from 100% to 250% concerning the base composite [133].

Plant extracts with recognized wood preservation qualities were used to integrate metal NPs [134]. The synergistic impacts of plant extracts' distinctive properties and the biocidal properties of metal NPs were exploited. Copper oxide NPs were prepared utilizing leaf concentrates of Neem (*Azadirachta indica*), (*Pongamia pinnata*), (*Lantana camara*) and orange peel (*Citrus reticulata*). All plants have wood preservative properties [135, 136] and are accounted for in the combination of metal NPs [61, 137, 138].

CuO NPs blended with plant extract work as reducing and settling agents, according to available data [139, 140]. Many plant extracts have been found to have antifungal properties as well as wood preservation effects [141]. *Langana camara* has wood-protecting characteristics, as confirmed by its capacity to increase the dimensional strength of wood. CuO NPs and *L. camara* extricates have wood additive qualities; therefore, their combined effect would result in a compelling eco-friendly and more affordable wood additive [142].

### 7.3.2 Nanomaterials as Coating for Wood

Coating is the process of putting a layer on the surface of the substrate. Stains, varnishes and paints are examples of common wood coatings that can be used for both defensive and ornamental purposes. The principle segments of the coatings determine their crucial qualities, such as binders, pigments, solvents, fillers, and additives [143].

Nanotechnology might be used in wood coatings to provide additional upgrades or even to bring about new performance characteristics. Fillers, for example, silicate-based minerals, are usually utilized in coatings to expand the mechanical performance of a polymer matrix, especially the hardness and solidness, or reduce costs. When applied to micron-sized wood, they have two major drawbacks: they limit the material's flexibility and make the coating structure more difficult to understand [144, 145].

Nanocoating is a cycle in which a thin layer of approximately 100 nm thickness is stored on a substrate to improve specific qualities or provide new functionality. The use of nanocoating in wood and wood products improves the strength, mechanical characteristics, fire retardancy and ultraviolet retention as well as reduces water absorption. The expansion of NPs is one technique to improve the usability and end user benefit of nanocoating [146]. These NPs have exceptionally enormous surface-to-volume proportions because of their morphology, which permits them to connect intensively with their environmental factors, and their nanosize ensures straightforwardness is as yet supported [147–149].

Metal NPs are widely employed to accomplish these upgrades and can be fused to the coating formulation in two ways, through solution blending (physical approach) or in situ (chemical approach) [144, 150]. The main methodology entails expanding the polymer in a suitable dissolvable manner and applying pressures to create dispersion. Spraying, brushing, or dipping is used to apply the nanoassembled coating on a wood surface [151–154].

Coating based on ZnO nanoparticles provides a protective barrier against moisture on wood composites. Wood samples with ZnO nanoparticles changed coatings from two coniferous and deciduous wood species [155].

An aqueous acrylic fluid coating for wood with ZnO nanoparticles as an ultraviolet safeguard including 2 wt% was developed. Treated wood with this coating exhibited excellent hydrophobicity, as well as decent photostability and gloss retention [156].

Lignin was segregated from iroko wood residues containing sawdust and blended sawdust and Norway spruce (IRNS), which were provided by a local wood houses manufacturer [157]. Isolated lignin parts were then used in a nondissolvable method to synthesize LNPs. LNPs were then used in the dip coating procedure to treat the wood surface.

Furthermore, nanobased coatings with solid bacterial characteristics, including silver, zinc oxide, copper, and titanium dioxide NPs, can protect wood and wood products from decay and, as a result, reduce bacterial action. The controlled and moderate arrival of the dynamic fixing is significant when these kinds of materials are evaluated for usage in a coating treatment because of its drawn-out constructive outcome and its insignificant ecological effect [144].

## 7.4 Conclusions, Outlook, and Future Aspects

Wood-composite panels offer a homogeneous structure that can be used in many industrial, construction, artistic, and public applications. There are some limitations that make the application of wood-based composite materials difficult, including water absorption and consequent swelling, and vulnerability to fire.

Nanomaterials enhance the properties of the original material and show great compatibility with traditional materials. Thus, it requires collective, concert and intensive efforts from various players, including stakeholders, industry and researchers, to promote the use of nanomaterials that could enter the next level of application and commercialization. Based on earlier research, it has been found that nanoparticles could be a viable replacement for wood preservatives, wood treatments, coatings, and adhesives and may play an important role in the next generation of wood composite systems. Futuristic research in the development and optimization of NPs in wood-based composites will also increase the bonding between nanomaterials and the base polymer matrix.

The application of nanomaterials readily reduces these limitations and improves characteristics such as physical, mechanical, UV protection, hygroscopicity and resistance against mold, decay fungi, borers, and termites. In summary, it is concluded that during the last decade, many efforts have been made for the development of green synthesis. Green synthesis gives headway over chemical and physical methods, as it is cost-effective, eco-accommodating, and effectively scaled up for large-scale synthesis.

## References


1. Zhao, L.F., Liu, Y., Xu, Z.D., Zhang, Y.Z., Zhao, F. and Zhang, S.B., State of research and trends in development of wood adhesives, *Forestry Studies in China*, 2011, **13**, p. 321.
2. He, Z., Zhang, Y. and Wei, W., Formaldehyde and VOC emissions at different manufacturing stages of wood-based panels, *Building and Environment*, 2012, **47**, p. 197.
3. Böhm, M., Salem, M.Z.M. and Srba, J., Formaldehyde emission monitoring from a variety of solid wood, plywood, blockboard and flooring products manufactured for building and furnishing materials, *Journal of Hazardous Materials*, 2012, **221–222**, p. 68.
4. Roffael, E., Volatile organic compounds and formaldehyde in nature, wood and wood based panels, *Holz als Roh-und Werkstoff*, 2006, **64**, pp.144–149.
5. Heinrich, L.A., Future Opportunities for bio-based adhesives-advantages beyond renewability, *Green Chemistry*, 2019. **21**, pp. 1866–1888.
6. Jang, Y., Huang, J. and Li, K., A new formaldehyde-free wood adhesive from renewable materials, *International Journal of Adhesion and Adhesive*, 2011. **31**, pp. 754.
7. Prasittisopin, L. and Li, K., A new method of making particleboard with a formaldehyde-free soy-based adhesive, Composites Part A: *Applied Science and Manufacturing*, 2010. **41**, pp. 1447–1453.
8. Gadhave, R.V., Mahanwar, P.A. and Gadekar, P.T., Starch-based adhesives for wood/wood composite bonding: review. *Open Journal of Polymer Chemistry*, 2017. **7**, pp.19–32.
9. Gu, Y., Cheng, L., Gu, Z., Hong, Y., Li, Z. and Li, C., Preparation, characterization and properties of starch-based adhesive for wood-based panels, *International Journal of Biological Macromolecules*, 2019. **134**, pp. 247–254.
10. Jiang, Y., Chen, Q., Tan, H., Gu, J. and Zhang, Y., A low-cost, formaldehyde-free, and high-performance starch-based wood adhesive, *BioResources*, 2019. **14**, pp. 1405.
11. Wang, Z., Zhu, H., Huang, J., Ge, Z., Guo, J., Feng, X. and Xu Q., Improvement of the bonding properties of cassava starch-based wood adhesives by using different types of acrylic ester, *International Journal of Biological Macromolecules*, 2019. **126**, pp. 603–611.

12. Wang, Z., Li, Z., Gu, Z., Hong, Y. and Cheng, L., Preparation, characterization and properties of starch-based wood adhesive, *Carbohydrate Polymers*, 2012. **88**, pp. 699–706.

13. Vnučec, D., Kutnar, A. and Goršek, A., Soy-based adhesives for wood-bonding: a review, *Journal of Adhesion Science and Technology*, 2017. **31**, pp. 910–931

14. Xu, H.N.; Ma, S.; Lv, W.; Wang, Z., Soy protein adhesives improved by $SiO_2$ nanoparticles for plywoods, *Pigment Resin Technology*, 2011, **40**(3), pp. 191–195.

15. Kaboorani, A., Riedl, B., Blanchet, P., Fellin, M., Hosseinaei, O. and Wang, S., Nanocrystalline cellulose (NCC): a renewable nano-material for polyvinyl acetate (PVA) adhesive, *European Polymer Journal*, 2012. **48**, pp.1829–1837.

16. López-Suevos, F., Eyholzer, C., Bordeanu, N. and Richter, K., DMA analysis and wood bonding of PVAc latex reinforced with cellulose nanofibrils, *Cellulose*, 2010. **17**, pp. 387.

17. Islam, M.T., Alam, M.M., Patrucco, A., Montarsolo, A. and Zoccola, M., Preparation of nanocellulose: a review, *AATCC Journal of Research*, 2014. **1**, pp. 17–23.

18. Ferreira, F.V., Mariano, M., Rabelo, S.C., Gouveia, R.F. and Lona, L.M.F., Isolation and surface modification of cellulose nanocrystals from sugarcane bagasse waste: from a micro- to a nano-scale view, *Applied Surface Science*, 2018. **436**, pp. 1113–1122.

19. Mondal, S., Preparation, properties and applications of nanocellulosic materials, *Carbohydrate Polymers*, 2017. **163**, pp. 301–316.

20. Mesquita, R.G.A., Mendes, L.M., Sanadi, A.R., de Sena, N. and Alfredo, R., Urea formaldehyde and cellulose nanocrystals adhesive: studies applied to sugarcane bagasse particleboards, *Journal of Polymers and the Environment*, 2018. **26**, pp. 3040–3050.

21. Chen, H., Nair, S.S., Chauhan, P. and Yan, N., Lignin containing cellulose nanofibril application in pMDI wood adhesives for drastically improved gap-filling properties with robust bondline interfaces, *Chemical Engineering Journal*, 2019. **360**, pp. 393–401.

22. Scott, N. and Chen, H., Nanoscale science and engineering for agriculture and food systems, *Industrial Biotechnology*, 2013. **8**(6), pp. 340–343.

23. Akhtari, M. and Arefkhani, M., Application of nanotechnology in wood. In Proceedings of IRG Annual Meeting, IRG/WP 10–30542. Biarritz, France: *The International Research Group on Wood Protection*, 2010.

24. Clausen, C.A., Nanotechnology: Implications for the wood preservation Industry. In Proceedings of 38th Annual Meeting of IRG/WP 07-30415. Jackson Hole, WY: *The International Research Group on Wood Protection*, 2007.

25. Tarmian, A., Sepehr, A. and Gholamiyan, H., The use of nanosilver particles to determine the role of reverse temperature gradient in moisture flow in wood during lowintensity convective drying, *Special Topics and Reviews in Porous Media*, 2012. **3**(2), pp.149–156.

26. Nosal, E. and Reinprecht, L., Antibacterial and anti-mold efficiency of ZnO nanoparticles present in melamine-laminated surfaces of particleboards, *Bioresources*, 2017. **12**(4), pp. 7255–7267.

27. Tian, D., Hu, J., Bao, J., Chandra, R. P., Saddler, J. N. and Lu, C., Lignin valorization: lignin nanoparticles as high-value bioadditive for multifunctional nanocomposites, *Biotechnology for Biofuels* 2017, **10** (1), pp. 192–202.

28. Rajinipriya, M., Nagalakshmaiah, M., Robert, M. and Elkoun, S., Importance of agricultural and industrial waste in the field of nanocellulose and recent industrial developments of wood based nanocellulose: a review, *ACS Sustainable Chemistry and Engineering*, 2018, **6**(3), pp. 2807–2828.

29. Richter, A.P., Bharti, B., Armstrong, H.B., Brown, J.S., Plemmons, D., Paunov, V.N., Stoyanov, S.D. and Velev, O.D., Synthesis and characterization of biodegradable lignin nanoparticles with tunable surface properties, *Langmuir*, 2016, **32**(25), pp. 6468–6477.

30. Liu, Y., Laks, P. and Heiden, P. Controlled release of biocides in solid wood: I. Efficacy against brown rot wood decay fungus (Gloeophyllum trabeum), *Journal of Applied Polymer Science*, 2002. **86**(3), pp. 596–607.

31. Liu, Y., Laks, P., and Heiden, P., Controlled release of biocides in solid wood: II. Efficacy against Trametes versicolor and Gloeophyllum trabeum wood decay fungi, *Journal of Applied Polymer Science*, 2002. **86** (3), pp. 608–614.

32. Liu, Y., Laks, P., and Heiden, P., Controlled release of biocides in solid wood: III. Preparation and characterization of surfactant-free nanoparticles, *Journal of Applied Polymer Science*, 2002. **86**(3), pp. 615–621.

33. Taghiyari, H.R., Fire-retarding properties of nanosilver in solid woods, *Wood Science and Technology*, 2012. **46**(5), pp. 939–952.

34. Taghiyari, H.R. and Bibalan O.F., Effect of copper nanoparticles on permeability, physical, and mechanical properties of particleboard, *European Journal of Wood and Wood Product*, 2013. **71**(1), pp. 69–77.

35. Taghiyari, H.R., Enayati, A. and Gholamiyan, H., Effects of nanosilver impregnation on brittleness, physical and mechanical properties of heat-treated hardwoods, *Wood Science and Technology*, 2012. **47**(3), pp.1–14.

36. Rassam, G., Ghofrani, M., Taghiyari, H.R., Jamnani, B. and Khajeh, M.A. Mechanical performance and dimensional stability of nanosilver impregnated densified spruce wood, *European Journal of Wood and Wood Product*, 2012. **70**(5) pp. 595–600.

37. Taghiyari, H.R, Rangavar, H. and Bibalan, O.F., Effect of nanosilver on reduction of hotpressing time and improvement in physical and mechanical properties of particleboards, *BioResources*, 2011, **6**(4), pp. 4067–4075.

38. Tan, K.S. and Cheong, K.Y., Advances of Ag, Cu, and Ag–Cu alloy nanoparticles synthesized via chemical reduction route, *Journal of Nanoparticle Research*, 2013. **15**(4), pp. 1–29.

39. Hulkoti, N.I. and Taranathm, T.C., Biosynthesis of nanoparticles using microbes: a review, Colloids and Surface B: *Biointerfaces*, 2014. **121**, pp. 474–483.

40. Schrofel, A., Kratosova, G., Krautova, M., Dobrocka, E. and Vavra, I., Biosynthesis of gold nanoparticles using diatoms-silica-gold and EPS-gold bionanocomposite formation, *Journal of Nanoparticle Research*, 2011. **13**, pp. 3207–3216.

41. Singaravelu, G., Arockiamary, J.S., Kumar, G.V. and Govindaraju, K., A novel extracellular synthesis of monodisperse gold nanoparticles using marine alga, Sargassum wightii Greville. Colloids and Surface B: *Biointerfaces*, 2007. **57**, pp. 97–101.

42. Mittal, A.K., Chisti, Y. and Banerjee, U.C., Synthesis of metallic nanoparticles using plant extracts, *Biotechnology Advances*, 2013, **31**, pp. 346–356.

43. Vijayaraghavan, K. and Ashokkumar, T., Plant-mediated biosynthesis of metallic nanoparticles: a review of literature, factors affecting synthesis, characterization techniques and applications, *Journal of Environmental Chemical Engineering*, 2017, **5**, pp.4866–4883.

44. Iravani, S., Green synthesis of metal nanoparticles using plants, *Green Chemistry*, 2011. **13**, pp. 2638–2650.

45. Nosal, E. and Reinprecht, L., Preparation and application of silver and zinc oxide nanoparticles in wood industry: the review, *Acta Facultatis Xylologiae Zvolen*, 2018. **60**(2), pp. 5–23.

46. Akharti, M., Kokandeh, M.G. and Taghiyari, H.R., Mechanical properties of Paulownia fortune wood impregnated with silver, copper and zinc oxide nanoparticles, *Journal of Tropical Forest Science*, 2012. **24**(4), pp. 507–511.

47. Amini, E., Azadfallah, M., Layeghi, M. and Taleri-Hassanloui, R., Silver-nanoparticle impregnated cellulose nanofiber coating for packaging paper, *Cellulose*, 2016. **23**, pp. 557–570.

48. Akarona, E., Koutzagioti, C., Salmas, C., Ntalos, G., Skoulikidou, M.C. and Tsamis, C., Enhancing wood resistance to humidity with nanostructured ZnO coatings, *Nano Structures and Nano Objects*, 2017. **10**, pp. 57–68.

49. Candan, Z., Environmentally friendly wood composites by nanocellulose. Madrid, Spain: COST FP1205, 2014.

50. Candan, Z., Gonultas, O. and Akbulut, T., Nanocellulose reinforced adhesives for wood composites. Goteborg, Sweden: COST FP1205, 2013.

51. Mulugeta H.W., Belingardi, G., Koricho, E.G. and Red, D.T., Effects of nanomaterials and particles on mechanical properties and fracture toughness of composite materials: a short review, *AIMS Materials Science*, 2019. **6**(6), pp. 1191–1212.

52. Hill, C.A.S., Wood modification: chemical, thermal and other processes. West Sussex, UK: John Wiley and Sons, 2006.

53. Rowell, R.M., Acetylation of wood: a review, *International Journal of Lignocellulosic Product*, 2014, **1**, pp. 1–27.

54. Mohammadlou, M., Maghsoudi, H. and Jafarizadeh-Malmri, H., A review on green silver nanoparticles based on plants: synthesis, potential of applications and eco-friendly approach, *International Food Research Journal*, 2016. **23**(2), pp. 446–463.

55. Moldovan, B., David, L., Achim, M., Clichici, S. and Filip, G. A., A green approach to phytomediated synthesis of silver nanoparticles using Sambucus nigra L. fruits extract and their antioxidant activity, *Journal of Molecular Liquids*, 2016. **221**, pp. 271–278.

56. Kumar, J.S., Kumar, S.V. and Kumar, R.S., Synthesis of zinc oxide nanoparticles using plant leaf extract against urinary tract infection pathogen, *Resource-Efficient Technologies*, 2017. **3**(4), pp. 459–465.

57. Narayanan, K. and Park, H.H., Antifungal activity of silver nanoparticles synthesized using turnip leaf extract (Brassica rapa L.) against wood rotting pathogens, *European Journal of Plant Pathology*, 2014. **140**, pp.185–192.

58. Gawade, V.V., Gavade, N.L., Shinde, H.M., Babar, S.B., Kadam, A.N. and Garadkar, K.M., Green synthesis of ZnO nanoparticles by using Calotropis procera leaves for the photodegradation of methyl orange, *Journal of Materials Science: Materials in Electronics*, 2017. **28** (18), pp. 14033–14039.

59. Rathanasamy, R., Thangasamy, P., Thangamuthu, R., Sampath, S. and Viswanathan, A., Green synthesis of ZnO nanoparticles using Carica papaya leaf extracts for photocatalytic and photovoltaic applications, *Journal of Material Science: Materials in Electronics*, 2017. **28**(14), pp. 10374–10381.

60. Agrawal, H., Kumar, V.S. and Rajeshkumar, S., A review on green synthesis of zinc oxide nanoparticles: *an eco-friendly approach, Resource-Efficient Technologies*, 2017. **3**(4), pp. 406–413.

61. Ansilin, S., Kavya Nair, J., Aswathy, C., Rama, V., Peter, J. and Jeyachynthaya Persis, J., Green synthesis and characterisation of copper oxide nanoparticles using Azadirachta indica (Neem) leaf aqueous extract, *Journal of Nanoscience Technology*, 2016. **2**(5), pp. 221–223.

62. Kharissova, O.V., Dias, H.V.R., Kharisov, B.I., Pérez, B.O. and Pérez, V.M.J., The greener synthesis of nanoparticles, Trends in Biotechnology, 2013. **31**(4), 240–248.

63. Lin, X., Wang, F., Kuga, S., Endo, T., Wu, M. and Wu, D., Eco-friendly synthesis and antibacterial activity of silver nanoparticles reduced by nanowood materials, Cellulose, 2014. **21**(4), pp. 2489–2496.

64. Ajitha, B., Reddy, Y.A.K. and Reddy, P.S., Green synthesis and characterization of silver nanoparticles using Lantana camara leaf extract, *Materials Science and Engineering*: C, 2015. **49**, pp. 373–381.

65. Ismail, M., Gul, S., Khan, M.A. and Khan, M.I., Plant mediated green synthesis of antimicrobial nanoparticles: a review on recent trends, *Reviews in Nanoscience and Nanotechnology*, 2016. **5**. (2), pp.119–135.

66. Lateef, A., Azeez, M.A., Asafa, T.B., Yakeen, T.A., Akinboro, A. and Oladipo, I.C., Biogenic synthesis of silver nanoparticles using a pod extract of Cola nitida: Antibacterial and antioxidant activities and application as a paint additive. *Journal of Tabiah University for Science*, 2016. **5**(2), pp. 551–562.

67. Fatimah, I., Pradita, Y. and Nurfalinda, A., Plant extract mediated of ZnO nanoparticles by using ethanol extract of Mimosa pudica leaves and coffee powder, *Procedia Engineering*, 2016. **148**, pp. 43–48.

68. Nagajyothi, P.C., Cha, S.J., Yang, I.J., Sreekanth, T.V.M., Kim, K.J. and Shin, H.M. Antioxidant and anti-inflammatory activities of zinc oxide nanoparticles synthesized using Polygala tenuifolia root extract, *Journal of photochemistry and Photobiology B: Biology*, 2015. **146**, pp. 10–17.

69. Qu, J., Yuan, X., Wang, X. and Shao, P., Zinc accumulation and synthesis of ZnO nanoparticles using Physalis alkekengi L, *Environmental Pollution*, 2011. **159**, pp. 1783–1788.

70. Karnan, T. and Selvakumar, S.A.S., Biosynthesis of ZnO nanoparticles using rambutan (Nephelium lappaceum L.) peel extract and their photocatalytic activity on methyl orange dye, *International Journal of molecular Structure*, 2016. **1125**, pp. 358–365.

71. Ramesh, P., Rajendran, A. and Subramanian, A., Synthesis of zinc oxide nanoparticle from fruit of Citrus aurantifolia by chemical and green method, *Asian Journal of Phytomedicine and Clinical Research*, 2014. **2**(4), pp.189–195.

72. Dobrucka, R. and Dlugaszewska, J., Biosynthesis and antibacterial activity of ZnO nanoparticles using Trifolium pratense flower extract, *Saudi Journal of Biological Sciences*, 2016. **23**(4), pp. 517–523.

73. Rajan, A., Cherian, E. and Baskar, G., Biosynthesis of zinc oxide nanoparticles using Aspergillus fumigatus JFC and its antibacterial activity. *International Journal of Modern Science and Technology*, 2016. **1**(2), pp. 52–57.

74. Selim, Y.A., Azb, M.A., Ragab, I. and Abd El-Azim, M.H.M., Green synthesis of zinc oxide nanoparticles using aqueous extract of Deverra tortuosa and their cytotoxic activities, *Scientific Reports*, 2020. **10**, pp. 3445.

75. Pai, G., Dayal, N., Shettigar, C.D., Patil, P. and Thakur, M., Microwave assisted biosynthesis of silver nanoparticles by aqueous extract of *Ocimum sanctum* (tulsi), *MGM Journal of Medical Sciences*, 2014, **1**(3), pp. 117–120.

76. Henríquez, L.C., Aguilar, K.A., Álvarez, J.U., Fernández, L.V., Vásquez, J.M.D. and Baudrit, J.R.V., Green synthesis of gold and silver nanoparticles from plant extracts and their possible applications as antimicrobial agents in the agricultural area, *Nanomaterials* 2020, **10**, pp.1763.

77. Lee, K.Y., Aitomaki, Y., Berglund, L.A., Oksman, K. and Bismarck, A., On the use of € nanocellulose as reinforcement in polymer matrix composites, *Composite Science and Technology*, 2014. **105**, pp.15–27.

78. Nechyporchuk, O., Belgacem, M.N. and Bras, J., Production of cellulose nanofibrils: a review of recent advances, *Industrial Crops and Product*, 2016. **93**, pp. 2–25.

79. Damásio, R.A.P., Carvalho, A.G., Gomes, F.J.B., Carneiro, A.de C.O., Ferreira, J.C. and Colodette, J.L., Effect of CNC interaction with urea-formaldehyde adhesive in bonded joints of *Eucalyptus* sp., *Forest Science*, 2017. **45**, pp. 169–176.

80. Veigel, S., Rathke, J., Weigl, M. and Gindl-Altmutter, W., Particle board and oriented strand board prepared with nanocellulose-reinforced adhesive, *Journal of Nanomaterials*, 2012. **2012**, pp. 1–8.

81. Kargarzadeh, H., Ioelovich, M., Ahmad, I., Thomas, S. and Dufresne, A., Methods for extraction of nanocellulose from various sources, *in Handbook of nanocellulose and cellulose nanocomposites*, Kargarzadeh, H., Ahmad, I., Thomas, S. and Dufresne A. (eds.), Weinheim: Wiley-VCH, 2017, pp. 1–49.

82. Kunjalukkal Padmanabhan, S., Esposito Corcione, C., Nisi, R., Maffezzoli, A., and Licciulli, A., Polydiethyleneglycol–bisallyl carbonate matrix transparent nanocomposites reinforced with bacterial cellulose microfibrils, *European Polymer Journal*, 2017. **93**, pp.192–199.

83. Gellerstedt, G., Pranda, J. and Lindfors, E.L., Structural and molecular properties of residual Birch Kraft Lignins, *Journal of Wood Chemistry and Technology*, 1994, **14**, pp. 467.

84. Zikeli, F., Vinciguerra, V., Taddei, A.R., D'Annibale, A., Romagnoli, M. and Mugnozza, G.S., Isolation and characterization of lignin from beech wood and chestnut sawdust for the preparation of lignin nanoparticles (LNPs) from wood industry side-streams, *Holzforschung*, 2018. **72**, pp. 961.

85. Lievonen, M., Valle-Delgado, J.J., Mattinen, M.L., Hult, E.L., Lintinen, K., Kostiainen, M.A., Paananen, A., Szilvay, G.R., Setälä, H. and Österberg, M., A simple process for lignin nanoparticle preparation, *Green Chemistry*, 2016. **18**, pp. 1416–1422.

86. Beisl, S., Friedl, A. and Miltner, A., Lignin from micro- to nanosize: applications. *International Journal of Molecular Science*, 2017. **18**, pp. 2367.

87. Clausen, C.A., Enhancing durability of wood-based composites with nanotechnology, in *Nanocelluloses: potential materials for advanced forest products proceedings of nanotechnology in wood composites symposium*, Cai, Z. and Niska, K.O. (Eds.). General Technical Report. FPL–GTR–218. Madison, WI: U.S. Department of Agriculture, Forest Service, Forest Products Laboratory, 2012, pp 8–12.

88. López-Suevos, F., Eyholzer, C., Bordeanu, N. and Richter, K., DMA analysis and wood bonding of PVAc latex reinforced with cellulose nanofibrils, *Cellulose*, 2010. **17**, pp. 387.

89. De Filpo, G., Palermo, A.M., Rachiele, F., Nicoletta, F.P., Preventing fungal growth in wood by titanium dioxide nanoparticles, *International Biodeterioration and Biodegradation*, 2013. **85**, pp. 217–222.

90. Cristea, M.V., Riedl, B. and Blanchet, P., Effect of addition of nanosized UV absorbers on the physico-mechanical and thermal properties of an exterior waterborne stain for wood, *Progress in Organic Coating*, **2011**. 72(4), pp. 755–762.

91. Francés Bueno, A.B., Bañón, N., Martínez de Morentín, L. and Moratalla García, J., Treatment of natural wood veneers with nano-oxides to improve their fire behavior, *Management Science and Engineering*, 2014. **64**(1), pp.12021.

92. Salma, U., Chen, N., Richter, D.L., Filson, P.B., Dawson-Andoh, B., Matuana, L., Amphiphilic core/shell nanoparticles to reduce biocide leaching from treated wood, 1-leaching and biological efficacy, *Macromolecular Material and Engineering*, 2010, **295** (5), pp. 442–450.

93. Peteu, S.F., Oancea, F., Sicuia, O.A., Constantinescu, F. and Dinu, S., *Responsive polymers for crop protection*, Polymers (Basel), 2010. **2**(3), pp. 229–251.

94. Moon, R.J., Frihart, C.R. and Wegner, T., Nanotechnology applications in the forest products industry, *Forest Product Journal*, 2006. **56** (5), pp. 4–10.

95. McCrank, J., *Nanotechnology applications in the forest sector*, Ottawa: Natural Resources Canada, 2009.

96. Taghiyari, H.R., Nanotechnology in wood and wood-composite materials, *Journal of Nanomaterials and Molecular Nanotechnology*, 2014, **3**, pp.1.

97. Jasmani, L., Rusli, R., Khadiran, T., Jalil, R. and Adnan, S., Application of nanotechnology in wood-based products industry: a review, *Nanoscale Research Letter*, 2020. **15**, pp. 207.

98. Nosal, E. and Reinprecht, L., Antibacterial and anti-mold efficiency of silver nanoparticles present in melamine-laminated particleboard surfaces, *BioResources*, 2019. **14**(2), 3914–3924.

99. Gao, W. and Du, G., Physico-mechanical properties of plywood bonded by nano cupric oxide (CuO) modified PF resins against subterranean termites, *Maderas Cienc Technology*, 2015. **17**(1), pp. 129–138.

100. Abd Norani, K., Hashim, R., Sulaiman, O., Hiziroglu, S., Ujang, S. and Noor, W., Biodegradation behavior of particleboard bonded with modifed PVOH/oil palm starch and nano silicon dioxide, *Iranian Journal of Energy and Environment*, 2017. **8**(4), pp. 269–273.

101. Silva, L.C.L., Lima, F.O., Chahud, E., Christoforo, A.L., Lahr, F.A.R. and Favarim, H.R., Heat transfer and physical-mechanical properties analysis of particleboard produced with ZnO nanoparticles addition, *BioResources*, 2019. **14**(4), pp. 9904–9915.

102. Gupta, A., Kumar, A., Sharma, K.V. and Gupta, R., Application of high conductive nanoparticles to enhance the thermal and mechanical properties of wood composite. Materials Today, 2018. **5**(1), pp. 3143–3149.

103. Muñoz, F. and Moya, R., Effect of nanoclay-treated UF resin on the physical and mechanical properties of plywood manufactured with wood from tropical fast growth plantations, *Maderas Cienc Technology*, 2018. **20** (1), pp.11–24.

104. Wang, X., Wang, S., Xie, X., Zhao, L., Deng, Y. and Li, Y., Multiscale evaluation of the effects of nanoclay on the mechanical properties of wood/phenol formaldehyde bondlines, *International Journal of Adhesive and Adhesion*, 2017. **74**, pp. 92–99.

105. Wibowo, E.S., Lubis, M.A.R., Park, B.D., Kim, J.S. and Causin, V., Converting crystalline thermosetting urea–formaldehyde resins to amorphous polymer using modified nanoclay, *Journal of Industrial and Engineering Chemistry*, 2020, **87**, pp.78–89.

106. Dorau, B., Arango, R. and Green, F., An investigation into the potential of ionic silver as a wood preservative, Woodframe Housing Durability and Disaster Issues, Las Vegas, Nevada, Madison, WI: *Forest Products Society*, 2004. pp. 133–145.

107. Rezei, V.T., Usefi, A. and Soltani, M., Wood protection by nanosilver against white rot, In *5th Symposium in Science and Technology*, Mashhad, Iran, May 12–17, 2011.

108. Moya, R., Berrocal, A., Rodriguez-Zuniga, A., Vega-Baudrit, J. and Noguera, S.C., Effect of silver nanoparticles on white-rot wood decay and some physical properties of three tropical wood species, *Wood and Fiber Science*, 2014. **46**(4), pp. 527–538.

109. Moya, R., Rodriguez-Zuniga, A., Berrocal, A. and Vega-Baudrit, J., Effect of silver nanoparticles synthesized with NPsAg-ethylene glycol ($C_2H_6O_2$) on brown decay fungi of nine tropical woods, *Journal of Nanoscience and Nanotechnology*, 2017.**17**, pp. 1–8.

110. Pařil, P., Baar, J., Čermák, P., Rademacher, P., Prucek, R. and Sivera, M., Antifungal effects of copper and silver nanoparticles against white and brown-rot fungi, *Journal of Material Science*, 2017. **52**(5), pp. 2720–2729.

111. Taghiyari, H.R., Moradi-Malek, B., Kookandeh, M.G., Bibalan and O.F., Effect of silver and copper nanoparticles in particleboard to control Trametes versicolor fungus, *International Biodeterioration and Biodegradation*, 2014, **94**, pp. 69–72.

112. Taghiyari, H.R. and Moradiyan, A., Effect of metal nanoparticles on hardness in particle board, *International Journal of Nano Dimensions*, 2014, **5**(4), pp. 379–386.

113. Reinprecht, L., Vidholdová, Z. and Kožienka, M., Decay inhibition of lime wood with zinc oxide nanoparticles in combination with acryl resin, *Acta Facultatis Xylologiae Zvolen*, 2015, **57**(1), pp. 43–52.

114. Akharti, M., Taghiyari, H.R. and Kokandeh, M.G., Effect of some metal nanoparticles on the spectroscopy analysis of Paulownia wood exposed to white rot fungus, European *Journal of Wood and Wood Products*, 2013. **71**(2), pp. 283–285.

115. Lykidis, C. Mantanis, G., Adamopoulos, S., Kalafata, K. and Arabatzis, I., Effect of nanosized zinc oxide and zinc borate impregnation on brown rot resistance of black pine (Pinus nigra L.) wood, *Wood Material Science and Engineering*, 2013, **8** (4), pp.241–243.

116. Harandi, D., Ahmadi, H., Achachluei, M.M., Comparison of $TiO_2$ and ZnO nanoparticles for the improvement of consolidated wood with polyvinyl butyral against white rot. *International Biodeterioration and Biodegradation*, 2016. **108**, pp. 142–148.

117. Farahani, M.R.M. and Banikarim, F., Effect of nano-zinc oxide on decay resistance of wood-plastic composites, *Bioresources*, 2013. **8**(4), pp. 5715–5720.

118. Habibadze, S., Asghar, O., Farahani, M.R.M., Mashkour, M., Effect of nano-ZnO on decay resistance and artificial weathering of wood polymer composite. *Journal of Nanomaterials and Molecular Nanotechnology*, 2014. **3**(3).

119. Marzbani, P., Afrouzi, Y.M. and Omidvar, A., The effect of nanozinc oxide on particleboard decay resistance, Maderas, Ciencia y tecnología, 2015. **17**(1), 63–68.

120. Reinprecht, L., Iždinský, J. and Vidholdová, Z., Biological resistance and application properties of particleboards containing nanozinc oxide, *Advances in Materials Science and Engineering*, 2018. **2018**, pp. 8.

121. Taghiyari, H.R. and Nouri, P., Efects of nanowollastonite on physical and mechanical properties of medium-density fberboard, Maderas Cienc Technology, 2015. **17**(4), pp. 833.

122. Taghiyari, H.R. and Commons, L.C., Effects of adding nanowollastonite, date palm prunings and two types of resins on the physical and mechanical properties of medium-density fibreboard (MDF) made from wood fibers, Bois Forêts des Trop, 2018. **335**, pp. 49.

123. da Silva, A.P.S., Ferreira, B.S., Favarim, H.R., Silva, M.F.F., Silva, J.V.F., dos Anjos, A.M., Physical properties of medium density fberboard produced with the addition of ZnO nanoparticles, BioResources, 2019. **14**(1), pp.1618–1625.

124. Candan, Z. and Akbulut, T., Physical and mechanical properties of nanoreinforced particleboard composites, Maderas Cienc Technology, 2015. **17**(2), pp. 319–334.

125. Taghiyari, H.R. and Norton, J., Effect of silver nanoparticles on hardness in medium-density fiberboard (MDF), *iForest Biogeosciences and Forestry*, 2014. **8**(5), pp. 677.

126. Kawalerczyk, J., Dziurka, D., Mirski, R. and Szentner, K., Properties of plywood produced with urea-formaldehyde adhesive modified with nanocellulose and microcellulose, Drvna Industrija, 2020. **71**(1), pp. 61–67.

127. Hansted, F.A.S., Hansted, A.L.S. and Padilha, E.R.D., Caraschi, J.C. and Goveia, D. and de Campos, C.I., The use of nanocellulose in the production of medium density particleboard panels and the modification of its physical properties, BioResources, 2019. **14**(3), pp. 5071–5079.

128. Amini, E., Tajvidi, M., Gardner, D.J. and Bousfeld, D.W., Utilization of cellulose nanofbrils as a binder for particleboard manufacture, BioResources, **12**(2), pp. 4093–4110.

129. Hunt, J.F., Leng, W. and Tajvidi, M., Vertical density profle and internal bond strength of wet-formed particleboard bonded with cellulose nanofibrils, *Wood and Fiber Science*, **49**(4), pp. 1–11.

130. Leng, W., Hunt, J.F. and Tajvidi, M., *S*crew and nail withdrawal strength and water soak properties of wet-formed cellulose nanofbrils bonded particleboard, BioResources, 2017. **12**(4), pp. 7692–7710.

131. Antonelli, F., Galotta, G., Sidoti, G., Zikeli, F., Nisi, R., Petriaggi, B.D. and Romagnoli, M., Cellulose and lignin nano-scale consolidants for waterlogged archaeological wood, *Frontiers in Chemistry*, 2020. **8** (32), pp.1-32.

132. Herrera, M.A., Sirviö, J.A., Mathew, A.P. and Oksman, K., Environmentally friendly and sustainable gas barrier on porous materials: Nanocellulose coatings prepared using spin- and dip-coating, *Materials and Design*, 2016. **93**, pp. 19–25.

133. Eesaee, M. and Shojaei, A., Effect of nanoclays on the mechanical properties and durability of novolac phenolic resin/woven glass fiber composite at various chemical environments, *Composite Part A: Applied Science.* 2014. **63**, pp. 149–158.

134. Shiny, K.S., Sundararaj, R., Mamatha, N. and Lingappa, B., A new approach to wood protection: preliminary study of biologically synthesized copper oxide nanoparticle formulation as an environmentally friendly wood protectant against decay fungi and termites, *Maderas Ciencia y tecnología*, 2019. **21**(3), pp. 347–356.

135. Sotannde, O.A., Yager, G.O., Zira, B.D. and Usman, A., Termiticidal effect of neem extracts on the wood of Khaya senegalensis, *Research Journal of Forestry*, 2011. **5** (3), pp. 128–138.

136. Machado, G.O., Cookson, L.J., Christoforo, A.L., Polito, W.L., Silva, M.R., Calil, C. and Lahr, F.A.R., Wood preservation based on Neem oil: Evaluation of fungicidal and termiticidal Effectiveness, *Forest Product Journal*, 2013. **63**, pp. 202–206.

137. Sundrarajan, M., Ambika, S. and Bharathi, K., Plant-extract mediated synthesis of ZnO nanoparticles using Pongamia pinnata and their activity against pathogenic bacteria, *Advanced Powder Technology*, 2015. **26**, pp.1294–1299.

138. Majumder, D.R., Bioremediation: Copper Nanoparticles from Electronic-waste, *International Journal of Engineering Science and Technology*, 2012. **4**(10), pp. 4388–4389.

139. Lee, H.J., Lee, G., Jang, N.R., Yun, J.H., Song. J.Y. and Kim, B.S., Biological synthesis of copper nanoparticles using plant extract, *Nanotechnology*, 2011. **1**(1), pp. 371.

140. Khan, S.A., Shahid, S., Sajid, M.R., Noreen, F. and Kanwal, S., Biogenic synthesis of CuO nanoparticles and their biomedical applications: a current review, *International Journal of Advanced Research*, 2017. **5**(6), pp. 925–946.

141. Tascioglu, C., Mesut, Y., Selim, S. and Caglar, A., Antifungal properties of some plant extracts used as wood preservatives, *International Biodeterioration and Biodegradation,* 2013. **85**(1), pp. 23–28.

142. Gupta, H. Sharma, K.R. and Sharma, J.N., Fungal inhibition in wood treated with *Lantana camara L.* extract, in *Wood is good*, Pandey, K., Ramakantha, V., Chauhan, S., Kumar, A. (eds.), Singapore: Springer, 2017, pp. 269–276.

143. Sandberg, D., Additives in wood products—today and future development, in *Environmental impacts of traditional and innovative forest-based bioproducts*, London: Springer, 2016, pp 105–172.

144. Nikolic, M., Lawther, J.M. and Sanadi, A.R., Use of nanofilers in wood coating: a scientific review, *Journal of Coating Technology and Research*, 2015, **12**, pp. 445–461.

145. Havrlik, M. and Pyranova, P., Protection of wooden materials against biological attack by using nanotechnology, Acta Polytechnika 2015, **55**, pp. 101–108.

146. Hincapié, I., Künniger, T., Hischier, R., Cervellati, D., Nowack, B. and Som, C., Nanoparticles in facade coatings: a survey of industrial experts on functional and environmental benefts and challenges, *Journal of Nanoparticle Research*, 2015. **17**(7), pp. 287.

147. Künniger, T., Heeb, M., Arnold, M., *A*ntimicrobial efficacy of silver nanoparticles in transparent wood coatings*, European Journal of Wood Protection*, 2014. **72**(2), pp. 285–288.

148. Landry, V., Blanchet, P., Boivin, G., Bouffard, J.F. and Vlad, M., *UV-LED curing efficiency of wood coatings*, Coatings, 2015. **5**(4), pp. 1019–1033.

149. Iždinsky, J., Reinprecht, L., Nosal, E., Vidholdova, Z., Krokosova, J., The activity of bacteria on surfaces of wooden composites painted with acrylate coating with addition of silver nanoparticles, In *Understanding wood modification through an integrated scientific and environmental impact approach (ModWoodLife), COST Action FP 1407 3rd Conference "Wood Modification Research and Applications"*, Kuchl–Salzburg, Austria: Book of Abstracts, 2017, pp. 130–131.

150. Teng, T., Arip, M., Sudesh, K. and Lee, H., Conventional technology and nanotechnology in wood preservation: a review, BioResources, 2018, **13**, pp. 9220–9252.

151. Mantanis, G. and Papadopoulos, A.N., The sorption of water vapor of wood treated with a nanotechnology compound, *Wood Science and Technology*, 2010, **44**, pp. 515–522.

152. Zhang, Z., MacMullen, J., Dhakal, H.N., Radulovic, J., Herodotou, C. Totomis, M. and Bennett, N., Biofouling resistance of titanium dioxide and zinc oxide nanoparticulate silane/siloxane exterior facade treatments, *Built Environment*, 2013. **59**, pp. 47–55.

153. Ghaemy, M. and Bekhradnia, S., Thermal and photocuring of an acrylate-based coating resin reinforced with nanosilica, *Journal of Coating Technology and Research*, 2012. **9**, pp. 569–578.

154. Goffredo, G.B., Citterio, B., Biavasco, F., Stazi, F., Barkeli, S. and Munafo, P., Nanotechnology on wood: the effect of photocatalytic nano coatings against *Aspergillus niger, Journal of Cultural Heritage*, 2017. **27**, pp.125–136.

155. Makarona, E., Koutzagioti, C., Salmas, C., Ntalos, G., Skoulikidou, M. C. and Tsamis, C., Enhancing wood resistance to humidity with nanostructured ZnO coatings, *Nano-Structures and Nano-Objects*, 2017. **10**, pp. 57–68.

156. Cristea, M.V., Riedl, B. and Blanchet, P., *E*nhancing the performance of exterior waterborne coatings for wood inorganic nanosized UV absorbers, Progress in Organic Coatings, 2010. **69**(4), pp. 432–441.

157. Zikeli, F., Vinciguerra, V., D'Annibale, A., Capitani, D., Romagnoli, M. and Mugnozza, G.S., Preparation of lignin nanoparticles from wood waste for wood surface treatment, *Nanomaterials*. 2019, **9**, pp. 281.

# Chapter 8

# Green Nanotechnology Research Avenue in Medicinal Biology

**Subhendu Chakroborty,[a] Pravati Panda,[b] M. V. B. Unnamatla,[c] Priyanka Chandra,[d] and Victor Varela-Guerreroc[e]**

[a]*Department of Basic Sciences, IITM, IES University, Bhopal, Madhya Pradesh, India*
[b]*Department of Chemistry, Ravenshaw University, Cuttack, Odisha, India*
[c]*Universidad autonoma del estado de Mexico, Toluca de Lerdo, Mexico*
[d]*Department of Pharmaceutical Sciences and Technology, Birla Institute of Technology, Mesra, Ranchi, India*
[e]*Centro Conjunto de Investigación en Química Sustentable UAEM-UNAM, México*
subhendu.cy@gmail.com

Nature has always surprised chemists and biologists through its natural resources, which can be used to treat various life-threatening diseases since ancient times. In the Ramayan, also, with the help of natural therapy, that is, sanjivani, Lord Ram saved the life of his beloved younger brother Laxman. Taking inspiration from the Vedas, the scientific community is working day and night to explore nature and its sources increasingly. Currently, nanotechnology is used in every aspect of life, starting from makeup to life-saving drug molecules. Due to its smallest particle size and high surface-

*Green Nanoarchitectonics*
Edited by Kaushik Pal

ISBN 978-981-4968-50-8 (Hardcover), 978-1-003-31860-6 (eBook)
www.jennystanford.com

to-volume ratio, its synthesis and application have become frontier research topics in recent years. However, nanoparticle synthesis using natural resources and further applying it for the manufacture of life-saving drug molecules is one of the current trending topics, that is, nanobiotechnology. Hence, most research persons from various academia and pharmaceutical companies are curious to work in this field. Plants have a hidden tendency to remodel inorganic metal ions into nanoparticles through their gifted organic resources, which opens up a new door to study biochemical analysis. Pages of the literature are already enriched with the glorious jobs such as tissue engineering, detection of proteins and pathogens, and most importantly in as a drug carrier in drug delivery, and so on., offered by various nanoparticles (Ag, Au, Fe, Cu, Zn, Cd, Ti, Zr, etc.). Cost-effectiveness, low toxicity, easy route of synthesis, production of less hazardous waste, environmental friendliness, and so onetc., are tremendous versatile properties that are associated with the green strategies of nanoparticle synthesis and thus make it a unique desirable technique over conventional methods of synthesis. In this regard, we think this chapter will be beneficial for research to gather more information regarding the synthesis of nanoparticles following greener protocols and their diversified application in various medicinal fields, such as antibacterial, anticancer, antioxidant, antiviral, and antifungal applications.

## 8.1 Green Chemistry Approach to Medicinal Nanoscience and Nanotechnology: An Overview

Nanotechnology is one of the most advanced science research fields and has received a great deal of attention among researchers worldwide. Nanoparticles (NPs) are generally small in size (below <100 nm) and thereby possess a large surface-to-volume ratio. Due to its small particle size, it can easily penetrate inside the cell membrane and attach to the target of interest. Hence, it possesses a wide range of applicability in medicinal chemistry, biomedicine, drug delivery systems, and so on. Among the different types of nanomaterials, metal (Ag, Zn, Cu, Fe, Au)- and metal oxide (ZnO, CuO,

$TiO_2$, $Y_2O_3$, CdO, ZrO)-based nanoparticle syntheses are of particular interest and have been studied extensively because they possess significant biological properties, such as antibacterial, anticancer, antioxidant, antifungal, anti-inflammatory, and antiviral properties. NPs can be synthesized by the following three different techniques: chemical, physical, and biological processes. The physical and chemical method of preparation is termed the conventional synthesis, also known as a bottom-up approach, in which a group of atoms/molecules are used for the synthesis of large materials/particles and the biological process of the synthesis of NPs is called the green synthesis/top-down approach, where a large material of interest is reduced to a small size to obtain nanostructures. Although the conventional method of NP preparation has its own merits, it suffers from limitations such as the high cost of reagents, hazardous chemicals, toxicity, and high temperature. This is harmful to the environment and human health. Hence, an ultimate and environmentally friendly synthesis of NPs involves green synthesis, which avoids toxic chemicals and reagents. Green synthesis uses different parts of plant extracts, microorganisms, and enzymes to synthesize NPs that are safe to handle, less toxic, and do not require high temperature. The phytochemicals present in the plant extract/enzymes/microorganisms function as capping, reducing, and stabilizing agents for the synthesized NPs. The green synthesis process using different plant extracts/enzymes/microorganisms is gaining popularity daily because the synthesized NPs exhibit a vast number of biological properties, such as antibacterial, anticancer, antioxidant, antifungal, anti-inflammatory, and antiviral properties, and the particle size of the NPs can be controlled. The synthesis of nanoparticles through the green process is called nanobiotechnology, which has gained vast applicability for biomedical purposes [1–8].

This chapter summarizes the recent advancements in the preparation of different metal/metal oxide-based NPs following the green process (plant extracts/microorganisms) and their different potential biological applications. We believe that this study will help researchers and academicians gain ideas about the synthesis of various bioactive nanoparticles with a wide range of potential against a diverse range of life-killing diseases with minimum particle sizes through a greener protocol.

## 8.2 Green Synthesis of Metal Nanoparticle: Analysis and Biological Impacts

On the basis of the application of NPs in every sector of life, their synthesis following a greener protocol and then evaluating their biological properties are now the major area of research interest among scientists worldwide. Avoidance of toxic chemicals, low cost, and easy preparation method are highlighted features found in green synthesis.

The aqueous form of *Rumex obtusifolius* leaf extracts was used for the preparation of Ag NPs from $AgNO_3$ by optimizing various reaction conditions, such as the ratio of conc. of reactants, pH, and temperature. The experiment was conducted at room temperature and pressure, and the biogenic materials of the extract helped reduce Ag to Ag NPs. The development of Ag NPs was identified by the change in the color of the $AgNO_3$ solution. The prepared NPs were successfully characterized through ultraviolet-visible (UV-vis), scanning electron microscopy (SEM), Fourier transform infrared (FTIR), transmission electron microscopy (TEM), energy-dispersive spectroscopy (EDS), and X-ray diffraction (XRD). The bactericidal inhibitory properties of Ag NPs were tested against *B. cereus*, *S. aureus*, *E. coli*, and *S. typhii*. The NPs were found to possess good antibacterial potency with zones of inhibition of 8.0 mm against *S. aureus*, 6.4 mm against *B. cereus*, and 6.6 mm against *E. coli* [9].

Ag NPs were prepared from the seed extracts of *Pongamia pinnata*. It is a medicinal plant and has been associated with multiple functions, such as antibacterial, antidiabetic, antiulcer, and antineurodegenerative activities. The synthesized NPs were characterized by different spectroscopic analyses, such as XRD, UV–Vis, SEM, EDS, TEM, and FTIR analysis. NPs with particle sizes of 20–60 nm with spherical shapes and crystalline morphologies were confirmed by XRD, SEM, and TEM analysis. In addition to the synthesis, the authors also studied their multidirectional properties, such as antibacterial, antioxidant, and wound healing abilities. NPs showed excellent antioxidant potential with $IC_{50} = 33.33 \pm 1.23$ ppm. Similarly, the NPs were also found to possess maximum growth inhibition potential, as indicated by their zone of inhibition, that is, $25 \pm 1.5$ mm toward *S. aureus*, $15 \pm 0.6$ mm toward *B. subtilis*, $24 \pm 0.6$ mm toward *P. aeruginosa*, and $20 \pm 0.6$ mm toward *E. coli*. They also showed considerable wound healing abilities with $p < 0.05$ over 5% betadine [10].

Acacia Lignin was used to prepare Ag NPs, which were then characterized by XRD, UV–Vis, EDS, SEM, FTIR, and TEM. The formation of the NPs was confirmed by the appearance of an absorption peak at 410 nm in the UV–Vis spectra. NPs were found to have an fcc structure with spherical morphology and particle sizes ranging from 10 to 50 nm. Further evaluation of their in vitro cytotoxicity toward the MCF7 breast cancer cell line showed that Ag NPs at 5 and 50 μg/ml concentrations were toxic to the cell, while the lignin extract was nontoxic. Additionally, they show good catalytic potential for reducing *p*-nitrophenol and methylene blue dye with rate constants of 0.01525 and 0.00098 $min^{-1}$, respectively. Moreover, they also exhibited potent antibacterial properties toward both G+ and G– bacterial strains [11].

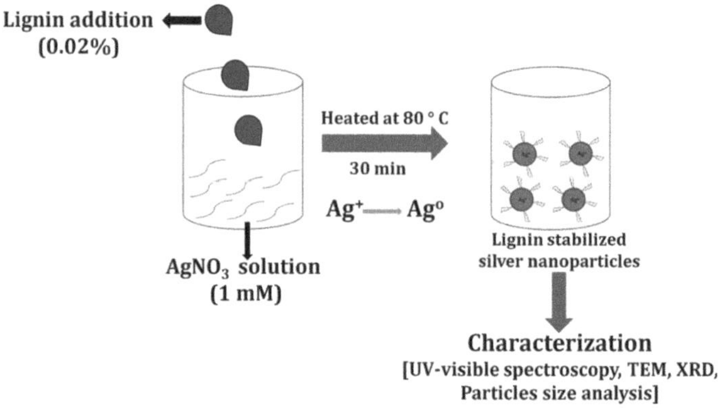


**Figure 8.1** Schematic representation of the preparation of AgNPs from Acacia lignin. Reprinted from Ref. 11, Copyright (2020), with permission from Elsevier.

Ag and Au NPs were synthesized from the aqueous peel extracts of *Passiflora edulis* by optimizing different reaction parameters to obtain a better colloidal solution. The biogenic materials present in the aqueous solution help in capping, reduction, and stabilization. The NPs were analyzed through UV–Vis, XRD, SEM, TEM, SAED, EDS, and FTIR. TEM data confirmed the spherical morphology with particle sizes of 7 nm and 25 nm for the Au and Ag NPs. The crystalline morphology of the NPs was identified by XRD and SAED data. The bactericidal potency of the NPs was screened against *B.*

*subtilis, E. coli*, and *S. aureus*. Ag NPs were found to exhibit strong bacterial inhibition potential against all tested pathogenic strains with an MIC value of 0.5 mM. NPs exhibit significant catalytic activity in the reduction and degradation of nitrophenols and dyes [12].

Cu NPs were synthesized from the plant extracts of *Cissus arnotiana,* and their in vitro antioxidant and antimicrobial activities were monitored. The formation of the nanoparticles was confirmed by further spectral analysis. The phytochemicals present in the plant extract perform multiple roles, such as capping, reducing, and stabilizing the synthesized NPs. The appearance of absorption peaks between 350 and 380 nm confirmed the formation of the NPs. Crystalline structures with spherical geometries possessing particle sizes between 60 and 90 nm were assessed through XRD, SEM, and TEM spectral studies. The surface roughness of the NPs was monitored through AFM. Biogenically prepared NPs displayed good inhibition against the G– (*E. coli*) bacterial strain, with a maximum zone of inhibition of 22.20 ± 0.16 mm at 75 μg/ml. NPs act as a fine antioxidant with potency equal to that of standard ascorbic acid at a concentration of 40 μg/ml [13].

Ag NPs were prepared from the extract of *Ferula persica* by optimizing the physiological parameters, that is, the concentration of extract and $AgNO_3$, temperature, pH, and incubation period. The syntheses of NPs were confirmed by UV–Vis, XRD, SEM, EDS, TEM, and FTIR analysis. Chlorogenic acid was found to be the essential content in the phenolic extract, which helps in the capping and reduction of Ag NPs, providing them with stability. The NPs were found to possess a homogeneous spherical shape with a size of 15 nm. Evaluation of the anticancer potential of the NPs was performed against two cancerous cell lines, that is, AGS (human gastric carcinoma) and MCF-7 (breast cancer), and it was found that the NPs exhibited good anticancer efficacy against both cancer cell lines with $IC_{50}$ values of 11.07 μg/ml and 21.28 μg/ml, respectively. Including the anticancer study, the NPs were tested for their bacterial inhibition abilities. The NPs showed good antibacterial potencies against *A. baumannii, P. aeruginosa*, and *S. aureus* at a concentration of 7.8 μg/ml. Additionally, the NPs exhibited good inhibitory activity against the leishmania parasite, with $IC_{50}$ values of 26.43 μg/ml for amastigotes and 23.14 μg/ml for promastigotes [14].

Ag NPs were prepared from the extracts of Tulsi and studied for their antibacterial activity. The structure, composition, and morphology of the NPs were confirmed through various spectroscopic analyses. The bacterial inhibitory activities were studied against G+ (*S. aureus*) and G– (*E. coli*) bacterial strains. The synthesized NPs exhibited good antibacterial potency [15].

Small-sized (2.9 ± 0.7 nm), monodispersed, and stable Ag NPs were synthesized by using soluble soybean polysaccharide (SSPS) via a greener protocol. In this process, the SSPS acts as a capping, stabilizing, and reducing agent. The prepared NPs were analyzed through UV–Vis, XRD, FTIR, EDX, SEM, and TEM techniques. The bactericidal potential was evaluated toward *S. aureus*, *E. coli*, and *P. aeruginosa*, and the results revealed that the NPs showed excellent bacterial inhibitory properties [16].

Green synthesis of Ag NPs was reported from lignin (an oil palm fruit branch extract) at varying concentrations. Lignin isolated under harsh reaction conditions (60 min, 3.5% acid) exhibited good antioxidant potential compared to those obtained under mild reaction conditions (20 min, 1.5% acid). The color change during synthesis indicated the formation of lignin-mediated Ag NPs. The average particle size of the obtained NPs was found to be between 18 and 20 nm with a polycrystalline morphology and spherical geometry. It showed a negative zeta potential in suspension. The synthesized NPs were subjected to an in vitro antibacterial study. The reports indicated that the NPs showed good antimicrobial properties with an MIC of 62.5 μg/ml [17].

Natural polysaccharides present in sweet lemon peel waste were used for the generation of Ag NPs. The carbon dot (CD) present in it helps in the stabilization and reduction of NPs at room temperature. The formation of CD and CD-doped Ag NPs was confirmed by UV–Vis, FTIR, SEM, XRD, TEM, and fluorescence spectrophotometry. The antibacterial and anticancer properties of the synthesized NPs were evaluated. The results obtained indicated that CD-doped Ag NPs showed prominent antimicrobial properties toward *E. coli* at a concentration of 5.0 μg/ml. However, compared to CD, CD-doped Ag NPs showed higher cytotoxicity toward the breast cancer (MCF-7) cell line at $IC_{50}$ = 10 μg/ml. Mechanistically, it was observed that the generation of reactive oxygen species causes cell death in MCF7 cells, suggesting that it can have further application in breast cancer treatment in the near future [18].

Cu NPs were prepared from the fruit extracts of *Citrus sinensis* under microwave irradiation and were analyzed through various spectroscopic techniques. A peak at 560 nm in the UV–Visible spectrum indicated the generation of Cu NPs. The biomolecules present in the orange juice extract helped in the bioreduction and stabilization of Cu NPs, as indicated by FTIR analysis. A globular shape with a smooth surface of Cu NPs was identified by SEM, and EDX spectra helped in the analysis of the elemental composition of NPs (Cu: 77.68%, C: 17.55%, and O: 4.77%). Morphologically round shapes with diameters of 6.93 to 20.70 nm were illustrated through TEM analysis. A high negative value of zeta potential (−25.60 mV) indicated the stability of the NPs. Additionally, the fcc crystal lattice geometry of NPs was confirmed through the XRD study. The synthesized NPs were screened for their in vitro anticancer and antibacterial abilities. The anticancer study was performed through an XTT cell viability assay, indicating that the NPs were nontoxic toward the L929 cell line (mouse fibroblast cell line). Additionally, the NPs displayed good bactericidal growth-inhibitory ability against *S. aureus* and *E. coli*. All these results implied that the synthesized NPs can further have scope in biomedical applications [19].

Ag NPs were synthesized through a convenient technique from extracts of three different *Sideritis* species, that is, SA (*S. argyrea*), SB (*S. brevidens*), and SL (*S. lycia*). *Sideritis* species were found to possess a wide range of photochemical properties, which helped reduce and stabilize the NPs. XRD suggested the fcc geometry of the NPs. The generation of the NPs was identified through an SPR study, where the corresponding peaks were found to resonate between 428 and 440 nm. TEM spectra confirmed the monodispersed and spherical nature of NPs with particle diameters between 22 and 26 nm. The presence of different functional moieties that help in the reduction of the NPs was confirmed through FTIR analysis. The synthesized NPs have been studied for their tyrosinase and cholinesterase (AChE and BChE) inhibitory abilities. The results indicated that the *Sideritis*-merged Ag NPs showed potent BChE inhibitory efficiency compared to AChE inhibition. SA combined with Ag NPs was an excellent inhibitor of tyrosine, with a concentration of 33.02 mg KAE (kojic acid equivalents)/g over SB and SL. The report suggested that the involvement of sideritis in green synthesis can open a new horizon for biotechnological applications [20].

An easy and convenient pathway for the synthesis of Au NPs was devised from aqueous leaf extracts of *Centaurea behen* at room temperature and was analyzed by various spectroscopic techniques. The crystalline nature with spherical geometry of NPs possessing diameter <50 nm was confirmed through XRD analysis. The in vitro cytotoxicity and antioxidant abilities of both Au NPs and the *C. behen* aqueous leaf extract were determined through MTT assays and DPPH assays. NPs exhibited promising cytotoxic effects toward the THP-1 (leukemia cancer cell line) cell line with an $IC_{50}$ = 25 μg/ml and the highest DPPH scavenging abilities of 31% and 14%, respectively [21].

Fe NPs were prepared from aqueous leaf extracts of *Chlorophytum comosum,* which helped in biogenic stabilization and reduction of the NPs. Different spectral analysis data identified the development of the NPs. XRD, SEM, and TEM analysis predicted an amorphous morphology and spherical shape with an average particle diameter of 100 nm. The in vitro antibacterial properties of the NPs were studied against G+ and G– bacterial strains, and impressive outcomes were found. The NPs were more effective toward the G+ (*S. aureus*) bacterial strain, with an MIC of 6 μg/ml [22].

Ag NPs were synthesized from the flower extracts of *Calendula officinalis* using $Ag_2SO_4$, and their bactericidal effect was studied against *Pectobacterium caratovorum*. The prepared NPs were analyzed through different spectral studies, such as XRD, UV–Vis, FTIR, SEM, and EDX. The appearance of an absorption maximum at 450 nm confirmed the formation of NPs with particle diameters of 18–24 nm. The antibacterial potential was determined against *Pectobacterium caratovorum*. The NPs had MIC and MBC values of 40 and 160 ppm, respectively. Additionally, they inhibited biofilm formation at 2.5 and 40 ppm, respectively [23].

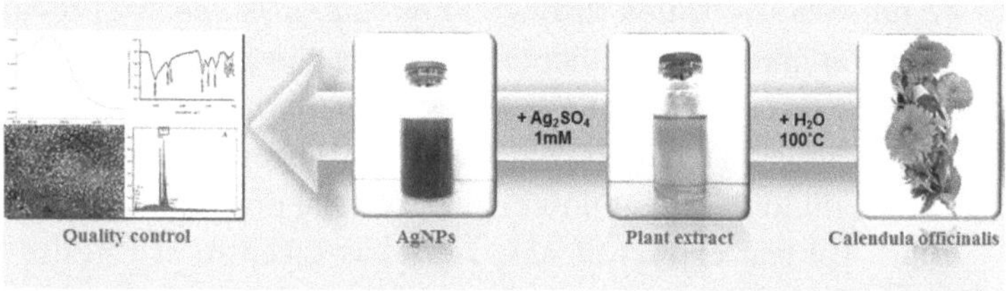


**Figure 8.2** Scheme involved in the green synthesis of silver NPs. Reprinted from Ref. 23, Copyright (2021), with permission from Elsevier.

Au NPs were prepared from curcumin and characterized through different spectral studies. The biomaterials present in

*Curcuma pseudomontana* play a vital role in the reduction of Au NPs. A peak resonating at 542 nm in the UV–Vis spectrum confirmed the development of NPs. Similarly, a homogenous structure with a spherical shape and particle diameter of 20 nm was predicted through XRD, SEM, and TEM data. The curcumin-doped Au NPs were subjected to screening for their in vitro antimicrobial potencies toward a large variety of pathogenic strains, such as *P. aeruginosa*, *B. subtilis*, *S. aureus*, and *E. coli,* and noticeable bacterial growth inhibition (94%) was observed. The NPs were also found to have greater antioxidant as well as radical scavenging abilities [24].

Ag NPs were prepared from the Chinese herbal extract of *Scutellaria barbata*, which helps in the reduction of $Ag^{+}$ to $Ag^{0}$, and were analyzed through UV–Vis, XRD, FTIR, TEM, and AFM. An absorption maximum at 400 nm confirmed the generation of Ag NPs. It was found to have a pure crystalline morphology and spherical shape with a particle diameter of 20–40 nm. The NPs were found to have potent antibacterial abilities with 60% inhibition of biofilm formation. The maximum zone of inhibition of the NPs was found to be 20–22 mm at a concentration of 60 mg/ml with an MIC of 2.2–3.1 µl [25].

*Syngonium podophyllum* (SP) dry leaf extracts in the aqueous form were used to prepare Ag NPs by mixing the extract solution with $AgNO_3$. The development of NPs was observed by the color change of the prepared sample mixture. The formed Ag NPs were analyzed perfectly through different spectroscopic studies. An absorption maximum at 420 nm indicated the generation of the NPs. The spherical shape with an average particle diameter of 2 to 47 nm was confirmed through SEM data. EDX spectra provided information about the elements present in the SP-capped Ag NPs. The antibacterial ability of the NPs was demonstrated through an agar well diffusion study against various pathogenic strains, and the NPs were found to exhibit potent antibacterial properties [26].

Inner gel leaf extract of *Aloe vera* was taken to synthesize Ag NPs and was characterized thoroughly. The color change and appearance of the absorption maxima at 400 nm indicated the generation of AgNPs. FTIR indicated different functional moieties (polysaccharides) in the extract, which helped in the capping, reduction and stabilization processes. The antimicrobial properties of the NPs were studied against both G+ (*S. aureus*) and G– (*P.*

*aeruginosa, E. coli,* and *Enterobacter sp.*) bacterial strains through the disc diffusion process. The combined effect of phytochemicals and Ag NPs during the capping process helps enhance bactericidal action against pathogens by reducing the particle size, which ultimately increases the surface-to-volume ratio of NPs [27].

Green syntheses of Ag/AgCl NPs were performed from the algal extract of *Chara vulgaris* living in fresh water using a 3:1 ratio of *Chara vulgaris* and $AgNO_3$, and their formation was confirmed through XRD, UV–Vis, and SEM techniques. The appearance of an absorption peak at 395 nm in the UV–VisVis spectrum confirmed the formation of NPs, and XRD revealed a fcc structure with the crystalline morphology of the generated NPs. Additionally, SEM confirmed that the average particle size of the NPs varied from 16.99 ± 0.3 nm. NPs were used to study their in vitro antibacterial abilities toward *E. coli, K. pneumonia, S. aureus,* and *P. aeruginosa* bacterial strains. The prepared NPs showed satisfactory antibacterial efficacy toward both G+ and G– bacterial strains. The combination of Ag/AgCl NPs with other antibiotics was found to significantly inhibit the resistance developed by harmful bacterial strains toward antibiotics [28].

Ag NPs were prepared from the gel extract of aloe vera, and their formation was confirmed by UV–Vis, XRD, EDX, SEM, and FTIR. Spectral data confirmed that NPs possess a crystalline structure and spherical morphology with a particle diameter of 40 nm. The obtained NPs exhibited attentive antibacterial properties against *E. coli, B. subtilis,* and *Pseudomonas* [29].

Ag NPs were prepared from *Bacillus subtilis* sp. *Suppress* (BSEPS) by optimizing the reaction parameters to obtain fine NPs, that is, the ratio of the concentration of the reactants (BSEPS and $AgNO_3$), pH, and temperature. The natural polysaccharide present in the solution helps in the reduction of the NPs. The as-prepared NPs were analyzed by UV–Vis, SEM, EDX, XRD, and FTIR spectroscopy. The prepared BSEPS/Ag NPs exhibited potent bacterial inhibition properties toward *E. coli* and *S. aureus* [30].

A convenient approach for Fe, Cu, and Ag NP preparation was developed by taking the leaf extracts of green and black tea and analyzed perfectly by different spectroscopic studies. The NPs were used to study their in vitro antibacterial properties toward vancomycin- and methicillin-resistant bacteria *S. aureus*. Additionally, the antifungal properties of the NPs were investigated against *A. parasiticus and Aspergillus flavus* fungal species. It was noticed that, compared to Cu and Fe NPs, Ag NPs exhibited remarkable antifungal and antibacterial properties [31].

Ag NPs were synthesized from the leaf extraction of *Melia azedarach, which acts as a capping, reducing, and stabilizing agent by using* $AgNO_3$*. The change in color of the reaction mixture from light yellow to brownish within 10 min indicated the formation of Ag NPs (*$Ag^+$ *was reduced to* $Ag^0$*). The as-prepared NPs were analyzed by various spectral analyses, such as XRD, FTIR, SEM, and EDX. The appearance of an absorption maximum at* 400 nm confirmed the generation of NPs. Different spectral studies identified a spherical shape with a fcc crystal lattice having an average particle size of 18–30 nm of NPs possessing a zeta potential of −13.1 mV. The in vivo and in vitro antifungal properties of Ag NPs were tested against *Verticillium dahlia. Ag NPs at a concentration of 20 ppm showed 97% growth inhibition in vivo toward the V. dahlia fungus and 87% growth inhibition in vitro at a concentration of* 60 ppm [32].

Zero-valent Ag NPs were obtained from *Bacillus* sp. strain and tested for their in vitro antifungal activity. Various spectral analyses identified the development of NPs. An absorption maximum at 447 nm confirmed the generation of the NPs. FTIR gave data about the capping of NPs by the bacterial proteins, which provided stabilization to the NPs. ZV-Ag NPs appeared as monodispersed spheres with crystalline surfaces, and particle sizes of 22.33–41.95 nm were confirmed through XRD, TEM, and SEM data. EDX confirmed the presence of Ag in 84.21% of the synthesized nanocomposite. The ZV-Ag NPs showed significant antifungal activity against *Colletotrichum falcatum* with conc. of 20 μg/ml by the damage of cellular organelles [33].

Ag NPs were synthesized from *Sporotrichum thermophile* extract using $AgNO_3$ following an easy, commercial, and environmentally friendly protocol by optimizing various reaction parameters. The color change of the reaction sample from transparent to dark brown signified Ag NPs by the biogenic reduction of $Ag^+$ to $Ag^0$. The synthesized NPs were characterized by different spectral analyses. The appearance of an absorption maximum at 426 nm revealed the development of Ag NPs with an average particle size of 40–70 nm, as confirmed by spectral data. The NPs showed prominent photocatalytic efficacy and reduced *p*-nitrophenol to *p*-aminophenol in the presence of $NaBH_4$ and light. Evaluation of the antibacterial properties of the NPs suggested that the NPs showed superior antibacterial activity toward G– bacterial strains compared to G+ because of their easy penetration inside the thin cell wall of G– bacteria, which consists of lipopolysaccharides [34].

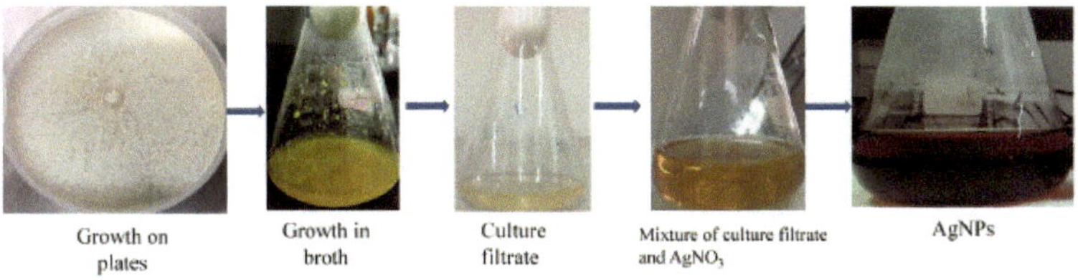


**Figure 8.3** Synthesis of Ag NPs from *S. thermophile* BJTLRMDU7. Reprinted from Ref. 34, Copyright (2020), with permission from Elsevier.

Ag NPs were prepared from the leaf extract of *Curcumae Kwangsiensis* Folium and were analyzed. Missing the characteristic band at 539 nm in the absorption spectrum due to $Ag^+$ marked the reduction of $Ag^+$ to $Ag^{0,}$ and the appearance of a band at 569 $cm^{-1}$ indicated the development of Au-O bonds. NPs possessed a sphere-like morphology with a particle diameter of ~8–25 nm, as confirmed by SEM data. After successful characterization, the NPs were used to study their in vitro antioxidant and anticancer (PA-1, SW-626, and SK-OV-3) properties. The NPs showed potent antioxidant activity by inhibiting DPPH at 153 μg/ml. Additionally, they were found to show good anticancer potential against ovarian cancer cells without affecting normal cells [35].

Ag NPs were synthesized from different plant extracts, such as *Crossopteryx febrifuga* (CR-FE), *Brilliantisia patula* (BR-PA), and *Senna siamea* (SE-SI). The as-prepared NPs were found to have spherical morphology with particle diameters of 45–110 nm. After confirmation of the formation of the NPs, they were studied for their in vitro antibacterial properties. Compared to the NPs extracted from crude and $AgNO_3$, NPs obtained from CR-FE, BR-PA, and SE-SI demonstrated superior antibacterial action against pathogenic strains [36].

Ag NPs were synthesized from the plant extracts of *Coriander sativum* and were well characterized by various spectral analyses. The appearance of absorption maxima at 400 nm confirmed the formation of NPs. The prepared NPs exhibited potential anticancer, antibacterial (zone of inhibition: G+: 11–13 mm, G–: 9–10 mm), and antioxidant properties [37].

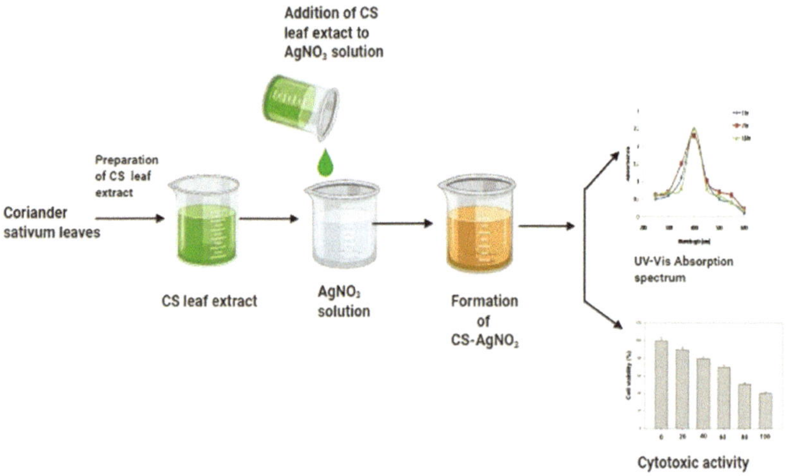


**Figure 8.4** Synthesis and biological activity evaluation of CS-AgNPs. Reprinted from Ref. 37, Copyright (2021), with permission from Elsevier.

Ag NPs were prepared from Ocimum canum Sims leaf extract and were characterized by XRD and SEM. XRD and SEM techniques with an average particle size of 15.72 nm proved a rod-like morphology with spherical shapes. The NPs showed potent antibacterial properties against G– (*E. coli*) bacteria with a zone of inhibition (ZOI) of 2.45 cm at 30 ppm [38].

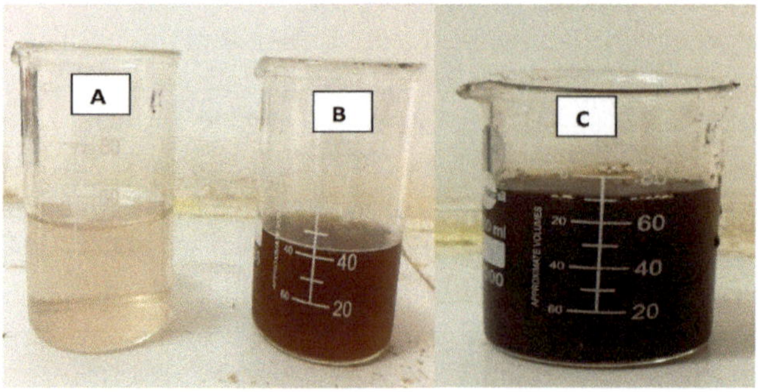


**Figure 8.5** (A) Aq. $AgNO_3$ (1 mM) solution. (B) Leaf extract of *Ocimum canum Sims*. (C) Color change of the solution. Reprinted from Ref. 38, Copyright (2020), with permission from Elsevier.

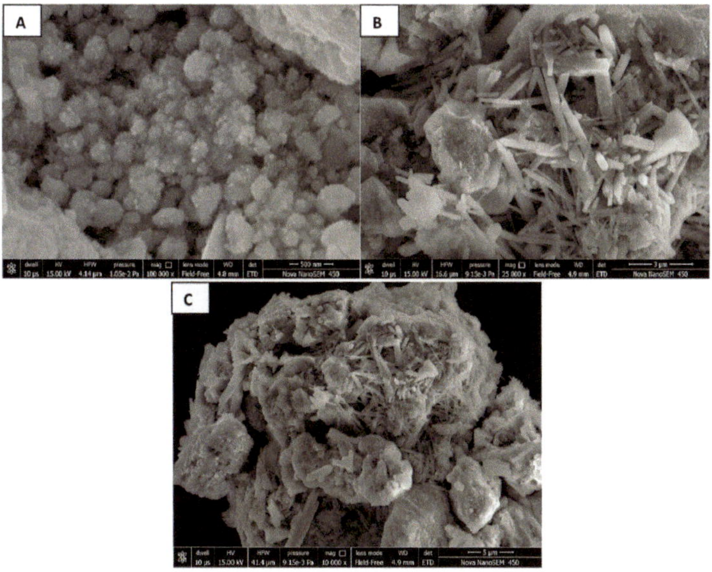


**Figure 8.6** SEM images of Ag NPs (A) 100,000× (B) 25,000× (C) 10,000× magnification. Reprinted from Ref. 38, Copyright (2020), with permission from Elsevier.

Ag NPs were synthesized from Carthamus tinctorius L. (safflower), and their antibacterial properties were analyzed against G+ (*S. aureus*) and G– (*P. fluorescens*) bacterial strains. NPs with a uniform spherical shape and crystalline nature with a size of 8.67 ± 4.7 nm were identified by spectral analysis. NPs were found to exhibit good antibacterial properties at a conc. of 0.9 μg/ml [39].

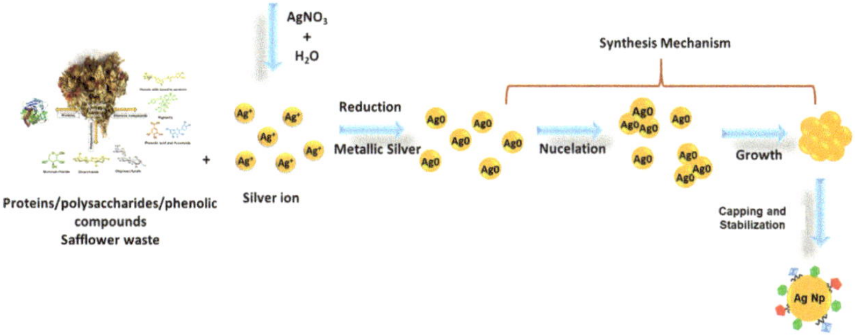


**Figure 8.7** Mechanism of Ag NP synthesis from aqueous safflower waste extract. Reprinted from Ref. 39, Copyright (2021), with permission from Elsevier.

Ag NPs were prepared from *P. niruri*, *A. paniculata*, and *T. cordifolia*, and their in vitro viral inhibition abilities toward chikungunya virus were evaluated. The formation, structure, morphology, and stability of the NPs were analyzed by different spectral analyses. NPs prepared from *A. paniculata* possessed a particle size of 70–95 nm, those synthesized from *P. niruri* had a particle diameter of 70–120 nm, and NPs prepared from *T. cordifolia* had an average particle size of 50–70 nm. The NPs exhibited good antiviral properties. Antiviral assay results also indicated that the inhibitory effect of NPs obtained from *A. paniculata* was found to be slightly higher, followed by *P. niruri* and *T. cordifolia* NPs [40].

Ag NPs were prepared from *Malephora lutea* F. and *Lampranthus coccineus*, and their in vitro antiviral activity was investigated by MTT assessment toward HAV-10, HSV-1, and Coxsackie B4 viral strains. The change in the color of the sample solution from yellow to reddish-brown indicated the generation of NPs by the biogenic reduction of Ag(I) ions to $Ag^{\circ}$, which was confirmed by different spectral studies. The spherical morphology with particle dimensions of 10.12–27.89 nm was determined by SEM and TEM studies. A peak at 417 nm in the absorption spectrum identified the generation of NPs. The prepared NPs were found to exhibit remarkable antiviral potencies against all the tested viral strains [41].

Ag NPs were synthesized from the bud extracts of *Syzygium aromaticum* and were studied for their in vitro antifungal potential against Newcastle viral disease (NDV) in embryonated chicken eggs. Different spectral analyses were performed to confirm the formation of NPs. An absorption peak at 450 nm in the absorption spectrum marked the generation of NPs by the reduction of $Ag^{+}$ to $Ag^{\circ}$. The particle diameter of the NPs was found to be 20 nm. The NPs were found to show good antiviral activity [42].

Ag NPs were synthesized from *Bacillus thuringiensis* using $AgNO_3$, and their viral inhibition properties toward plant viruses were studied. The prepared NPs were analyzed perfectly by different spectral analyses. The average particle diameter of the NPs was found to be between 10 and 20 nm. The absorption maxima at 450 nm confirmed the formation of NPs. The NPs showed good antiviral activity against *sunhemp rosette virus* at a concentration of 50 ppm [43].

Cu NPs were prepared from leaf extraction of *Celastrus paniculatus* and were characterized by different spectral analyses. The generation of NP was confirmed by the appearance of a peak at 269 nm. SEM and TEM data confirmed the spherical morphology with particle dimensions of 2–10 nm. FTIR gave information about the various functional moieties present in the prepared biogenic Cu NPs. A negative value of zeta potential (–22.2 mV) provided information about the extra stability of the NPs. The synthesized NPs showed good photocatalytic efficacy with a rate constant value of 0.0172 $min^{-1}$. Additionally, it exhibited good antifungal characteristics against *Fusarium oxysporum* (plant pathogenic fungi) with 76.29 ± 1.52% fungal inhibition [44].

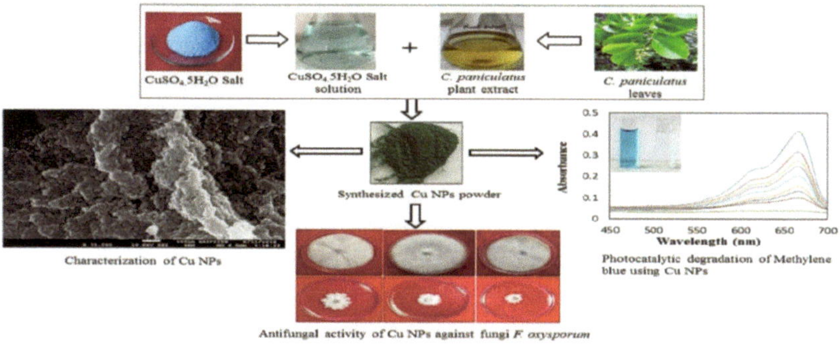


**Figure 8.8** Schematic representation for the synthesis of Cu NPs from the leaf extraction of *Celastrus paniculatus* Reprinted from Ref. 44, Copyright (2020), with permission from Elsevier.

Fe NPs were prepared from the leaf extract of *Calotropis procera* and were characterized perfectly. NPs were found to have an orthorhombic form with particle sizes ~32 nm. The prepared NPs were tested for their in vitro antifungal activity against *Alternaria alternata*. Fe NPs showed excellent effects and inhibited 87.9% of fungal growth at a concentration of 1.0 mg/ml [45].

## 8.3 Green Synthesis of Metal Oxide NPs and Their Biological Properties

Metal oxide NPs have vast applicability, such as in cosmetics, catalysts, semiconductors, drugs, optics, sensors, and so on. The wide demand for metal oxide nanoparticles in every aspect of life makes them unique and always encourages material chemists to think more about them and develop greener strategies for their preparation and application.

ZnO NPs were prepared from *Raphanus sativus var. Longipinnatus* leaf extracts using $(CH_3CO_2)_2Zn$ were analyzed through different techniques. The appearance of an absorption maximum at 369 nm in the UV–Vis spectrum indicated the generation of ZnO NPs. The synthesized NPs were found to have a wurzite crystal lattice, spherical shape, particle size of 209 nm and –13.7 mV zeta potential. The anticancer abilities of the NPs were studied against the A549 (lung cancer) cell line, and they showed good anticancer activity with $IC_{50}$ = 40 mg/ml [46].

CuO NPs were prepared following an easy, inexpensive, and environmentally eco-friendly protocol from the plant extracts of *Mussaenda frondosa* and were analyzed thoroughly. The appearance of absorption maxima between 377 and 382 nm in the UV spectrum proved the generation of CuO NPs. XRD, SEM, and TEM data detected the formation of pure monoclinic crystal lattices with spherical geometry having a particle diameter of 2–10 nm. FTIR and EDX gave information about the presence of Cu-O bonds with other functional groups along with the elemental composition of NPs. DLS suggested the monodispersity and high stability of the formed NPs. The NPs were found to be associated with multiple applications, that is, antioxidant, anticancer, antibacterial, antidiabetic, and anti-inflammatory properties. Moreover, it can be used to degrade the carcinogenic ability of methyl blue dye [47].

ZnO NPs were prepared from *Cassia auriculata* leaf extracts by using $ZnSO_4$ and tested for their bacterial inhibition potential. The prepared NPs were analyzed by FTIR, UV–Vis, XRD, SEM, TEM, and so on. AA spherical shape with an average particle diameter of 20 to 30 nm was found through SEM and EDX analysis. They also exhibited an optical band gap of 3.3 eV. The NPs exhibited strong antibacterial efficacy toward *B. subtilis*, *K. pneumonia*, *P. aeruginosa* and *P. mirabilis*. The capping reagent helped in the slow release of $Zn^{2+,}$ thereby improving the antibacterial property of the NPs [48].

CuO NPs were synthesized from two different parts of the plant extracts of *Madhuca longifolia* and were successfully characterized through different spectral analyses. NPs were found to have crystalline structures and spherical geometries with 120 and 30 nm average particle diameters, as determined by a spectral study. After that, they were evaluated for their antibacterial potencies toward G+ and G– bacterial strains. NPs showed good bacterial growth inhibition against all the tested strains, with a maximum zone of inhibition ranging from 14.67 to 15.67 mm at a concentration of 30 mg/ml [49].

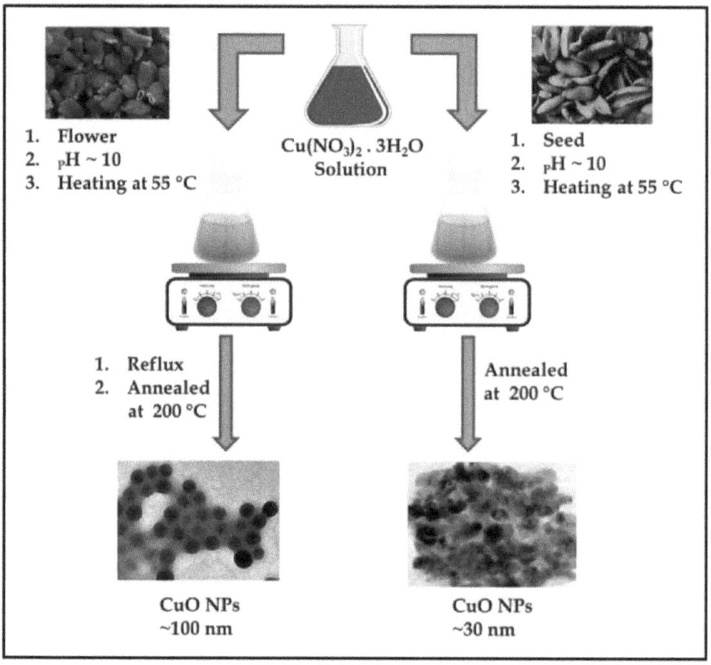


**Figure 8.9** Synthesis diagram of CuO NPs. Reprinted from Ref. 49, Copyright (2018), with permission from Elsevier.

ZnO NPs were prepared from leaf extracts of *Nilgiriantusciliantus* by using $ZnSO_4$ as the precursor under microwave irradiation and were successfully identified by XRD, SEM, UV–Vis, FTIR, and TEM analysis. The bacterial inhibition properties of ZnO NPs were studied in vitro against *S. mutans* and *S. aureus* (G+) and *P. aeruginosa* and *E. coli* (G–) bacterial strains via an agar well diffusion process. Additionally, the anticancer activity of the synthesized NPs was screened against MCF-7 (breast cancer cells) and L929 (normal fibroblast cells) cells through the MTT assay. The results obtained showed that the ZnO NPs showed good antibacterial activity and exhibited excellent anticancer potential. Delightedly, the anticancer study reports of the NPs signified that they were nontoxic toward normal cells, which suggests that the developed green technique of NP synthesis with promising biological properties can find applications for future biomedical research purposes [50].

A green protocol for the synthesis of CuO NPs was developed from *Ocimum tenuiflorum* by varying different concentrations of $Cu(CH_3CO_2)_2$. $H_2O$ at 70 °C. The synthesized CuO NPs were analyzed by XRD, UV–Vis, SEM, TEM, and EDS techniques. NPs with a crystalline nature and elliptical/spherical/rice-shaped morphology with a particle size of 6–18 nm were confirmed through SEM and TEM analysis. The antibacterial activity of the NPs was studied, which indicated that the NPs showed good growth inhibitory activities toward *E. coli*, *S. aureus*, and *B. subtilis*. In addition to the bacterial inhibitory property, the NPs also exhibited good photocatalytic efficiencies leading to 96.4 ± 0.83% methyl orange dye degradation in 24 min following pseudo first-order kinetics [51].

Ag–Au-decorated $Y_2O_3$ NPs were prepared from *Justicia adhatoda* leaf extraction using ionic liquid. [BMIM] $PF_6$ as the capping/stabilizing agent. The formation of NPs was analyzed through UV–Vis, XRD, SEM, EDX FTIR, and TEM analysis. XRD data confirmed that the NPs have a fcc crystal lattice with a particle diameter of 30 nm. NPs showed a Y–O stretching frequency band at 464–495 $cm^{-1}$ in the IR spectra and a 5.75 eV band gap, as confirmed by a UV-DRS study. EDX confirmed the presence of Ag, Au, Y, and O, and the flake-like morphology of the NPs was confirmed through SEM/TEM analysis. Screening of bacterial inhibition potential was carried out against *S. aureus* and *E. coli*, and the anticancer ability of the NPs was determined against the A549 (cervical cancer cell) cancerous cell line. The hybrid Ag–Au decorated $Y_2O_3$ NPs showed promising antibacterial and anticancer abilities compared to single $Y_2O_3$ NPs, Ag-doped $Y_2O_3$ NPs, and Au-doped $Y_2O_3$ NPs [52].

ZnO NPs were prepared from *Vitis vinifera* (grape extract), which acted as a natural reducing and stabilizing agent and was analyzed through various spectroscopic measurements. A wurzite-like crystalline structure with a Zn–O stretching band at 408.91 $cm^{-1}$ was confirmed through XRD and FTIR analysis. ZnO NPs were found to have significant bactericidal inhibition properties against G+ (*S. aureus*) and G– (*K. pneumoniae*) bacterial strains. Including this, it also showed potential cytotoxicity against AMGM5 and MCF-7-cell lines [53].

Synthesis of ZnO NPs was carried out from the leaf and plant extracts of *cannabis, aloe vera, J. curcas,* and *Tinosporacordifolia*. The morphology, crystallinity, and structure of the NPs were analyzed through FTIR, XRD, UV–Vis, SEM, EDX, and TEM studies. The NPs were found to have noteworthy bactericidal efficacy against both G+ and G– bacterial strains, that is, *K. pneumonia, S. aureus,* and *B. cereus* [54].

ZnO NPs were prepared from the fruit extraction of *Opuntia humifusa (a natural surfactant) involving a one-step hydrothermal process. The surfactant helped in the capping, reduction, and stabilization of the NPs. After successful synthesis, these were screened for their antimicrobial abilities. The NPs were found to have satisfactory inhibitory abilities toward multiple harmful bacterial pathogenic strains,* that is, *S. aureus, B. cereus, Pseudomonas, Proteus, and E. coli. The report suggested that the synthesized NPs can have multidirectional applicability from agriculture to medical purposes [55].*

CdO NPs were synthesized from turmeric plant extraction, which simultaneously acts as a natural stabilizing and reducing agent. The prepared NPs were characterized successfully through XRD, UV–Vis, SEM, FTIR, TEM, and AFM. CdO NPs were found to have a crystalline structure with a diameter of 33.31 nm and a 5.8 eV bandgap. The antibacterial properties of the NPs were studied against various pathogenic strains, and the obtained results proved that the CdO NPs were highly active toward *K. pneumoniae, P. aeruginosa, S. aureus,* and *E. coli*. Additionally, the NPs were examined for their anticancer abilities through the MTT assay. The NPs were found to possess excellent anticancer potential against HT29 (human colon cancer) cancerous cell lines [56].

ZnO NPs were processed from *Syzygium cumini* (Java plum) leaf extract, and their cytotoxicity and antioxidant properties were examined. The prepared NPs were found to possess moderate antioxidant potential and cytotoxicity against the A549 cell line [57]. Ag-doped ZnO NPs were prepared from *Morinda citrifolia* leaf extract through the coprecipitation method and were analyzed perfectly. FTIR identified that the biogenic materials present in the extract helped in the capping, stabilization, and reduction of the NPs. XRD and SEM confirmed that NPs possess hexagonal crystal lattices with spherical geometry. The prepared Ag-doped ZnO NPs exhibited 83% radical scavenging ability, confirming their antioxidant potential [58].

A cost-effective and convenient approach for the generation of ZnO NPs was reported from the bark extract of *Amygdalus scoparia* stem and then tested for its antibacterial, anticancer, antifungal, and antidiabetic properties. The morphology, elemental composition, functional groups, and structure of the NPs were determined by different spectral analyses. The results of the antibacterial study revealed that the synthesized NPs exhibited potential bacterial growth inhibition toward *E. coli*, *S. aureus*, *E. aerigenes*, *F. thapsinum*, *P. oryzae*, and *F. semitectum* compared to the standard antibiotics. The ZnO NPs also displayed potent cytotoxicity toward cancerous cells and were nontoxic toward normal cell lines. It also exhibited potent antidiabetic effects [59].

Fe-doped ZnO NPs were prepared from the leaf extracts of *Hibiscus rosa-sinensis and were analyzed thoroughly by using different spectroscopic techniques.* The NPs were found to have an average particle diameter between 15 and 170 nm. The developed Fe–ZnO NPs possessed excellent antibacterial, photocatalytic and antifungal properties compared to pure ZnO NPs [60].

A conventional protocol for preparing ZnO NPs was developed by using the peel extract of *Pithecellobium dulce* and was successfully characterized. The biogenic substances present in the peel extract act as capping, reducing, and stabilizing agents. The prepared NPs possessed a sphere-like geometry with a hexagonal crystal lattice. The antifungal activities of the NPs were studied, and the results obtained showed that the NPs inhibited the growth of *Aspergillus flavus* and *Aspergillus niger* in 37.81–63.57% at 500 and 1000 ppm, respectively. In addition to antifungal activity, the ZnO NPs also exhibited potent photocatalytic efficacy and promoted the degradation of methylene blue (63%) with a kinetic rate constant value of 0.00812 $min^{-1}$ [61].

A green methodology for the preparation of $TiO_2$ NPs was brought forward from the plant extracts of *Chenopodium quinoa, Trianthema portulacastrum and by a sol-gel (conventional) process. The formation of the NPs was confirmed by different spectral analyses, such as XRD, FTIR, SEM, and EDX. The average particle diameter of the*

*NPs synthesized from the three sources was found to be <15 nm. FTIR spectroscopy confirmed the presence of the Ti–O bond in the NPs in all three cases. The antifungal activity of the NPs was evaluated against wheat rust* (*Ustilago tritici*), a toxic plant pathogen. $TiO_2$ NPs obtained by both conventional and green processes showed good antifungal activity against *Ustilago tritici. However, superior antifungal activity was observed against wheat rust in NPs prepared from Chenopodium quinoa leaf extract [62].*

*Green synthesis of ZrO NPs was reported from the Enterobacter* sp. Their antifungal properties were studied against *P. versicolor (twig bright disease found in bayberry).* The development of the ZrO NPs was proven through XRD, UV–Vis, SEM, TEM, and FTIR studies. The presence of capping protein providing stabilization to the NPs was found through FTIR study. Similarly, NPs with a crystalline nature and spherical morphology with a particle diameter of 33–75 nm were confirmed by XRD, SEM and TEM data. An EDX study confirmed the elemental composition of the NPs (Zr: 54.40%, O: 43.49%). NPs exhibited considerable antifungal activity by the leakage of proteins and DNA with a zone of inhibition of 25.18 ± 1.52 mm at 20 μg/ml conc. toward *P. versicolor [63].*

ZnO NPs were prepared from different plant extracts, such as *Cinnamomum tamala, Beta vulgaris, Brassica oleracea* var. *Italica, Cinnamomum verum.* Various spectroscopic analyses, such as XRD, FTIR, SEM, TEM, and so on., were performed to confirm the formation of NPs. The antibacterial properties of the NPs were studied against G+ (*S. aureus*) and G– (*E. coli*) bacterial strains, and the results showed good bacterial growth inhibition efficiency of the NPs. However, ZnO NPs synthesized from *Beta vulgaris* were found to be ineffective against G+ bacterial strains, that is, *S. aureus*. The NPs were also subjected to in vitro antifungal activities toward *A. niger* and *C. albicans* fungal stains. NPs obtained from *Brassica oleracea* var. *Italica was found to be active against both fungal strains, whereas NPs obtained from C. tamala and B. vulgaris were* active toward *C. albicans and A. niger fungal strains [64].*

ZnO NPs were prepared from the flower extraction of *Nyctanthes arbor-tristis* using $(CH_3CO_2)_2Zn$. $2H_2O$ by the optimization of various reaction parameters to obtain fine nanocrystals and confirmed by different spectroscopic techniques. NPs were found to be highly stable and could be stored for four months in dried form. NPs with spherical shapes and crystalline morphologies with average particle diameters of 12–32 nm were confirmed by XRD, SEM, and TEM data. ZnO NPs were found to exhibit potential antifungal properties with an MIC of 16 μg/ml [65].

## 8.4 Conclusions, Outlook, and Future Aspects

In conclusion, this chapter summarizes the recent progress made in the synthesis of various metal and metal oxide-based NPs from different natural sources, such as plant extracts/microorganisms, and their diversified applications against various harmful diseases, such as antibacterial, anticancer, anti-inflammatory, antioxidant, antifungal, and antiviral activities. Green nanotechnology is becoming familiar day by day because it is safe to handle, cost-effective, and environmentally eco-friendly and ultimately provides natural capping, reducing and stabilizing agents. Additionally, fine nanoparticles with a size <100 nm can easily be synthesized by this process. In this chapter, the authors are planning to highlight the importance of natural agents/green nanotechnology for the synthesis of bioactive NPs. The main theme of this chapter is to avoid the use of hazardous chemicals as well as to avoid the less economic pathway of bioactive nanoparticle synthesis. In this chapter, the authors have also explained different views of the importance of NPs synthesized through the natural pathway and their further applications against different diseases. Therefore, the authors believe that this chapter will be beneficial for future researchers to plan and synthesize more fine nanoparticles following environmentally friendly and atom economic routes and to further check their applications against different harmful diseases.

### Conflicts of Interest

The authors declare no conflicts of interest.

### References

1. Teow, S.-Y., Wong, M. M.-T., Yap, H.-Y., Peh, S.-C., Shameli, K. *Molecules*, **2018**, *23*, 1366. DOI: 10.3390/molecules23061366.

2. Thacker, H. H., Ram, V. R., Dave, P. N. *Progress in Chemical and Biochemical Research*, **2019**, *2*, 84–91. DOI: 10.33945/SAMI/pcbr.2019.183239.1033.
3. Akbar, S., Tauseef, I., Subhan, F., Sultana, N., Khan, I., Ahmed, U., Haleem, K. S. *Inorganic and Nano-Metal Chemistry*, **2020**, *54*, 257–271. DOI: 10.1080/24701556.2019.1711121.
4. Waris, A., Din, M., Ali, A., Ali, M., Afridi, S., Baset, A., Khan, A. U. *Inorganic Chemistry Communications*, **2021**, *123*, 108369. DOI: 10.1016/j.inoche.2020.108369.
5. Vanlalveni, C., Lallianrawna, S., Biswas, A., Selvaraj, M., Changmai, B., Rokhum, S. L. *RSC Advances*, **2021**, *11*, 2804–2837. DOI: 10.1039/D0RA09941D.
6. Behzad, F., Naghib, S. M., Jadidikouhbanani, M. A., Tabatabaei, S. N., Zaree, Y., Rhee, K. Y. *Journal of Industrial and Engineering Chemistry*, **2021**, *94*, 92–104. DOI: 10.1016/j.jiec.2020.12.005.
7. Behravan, M., Panahi, A. H., Naghizadeh, A., Ziaee, M., Mahdavi, R., Mirzapour, A. *International Journal of Biological Macromolecules*, **2019**, *124*, 148–154. DOI: 10.1016/j.ijbiomac.2018.11.101.
8. Kumar, R., Roopan, S. M., Prabhakarn, A., Khanna, V. G., & Chakroborty, S. *Spectrochimica Acta Part A: Molecular and Biomolecular Spectroscopy*, **2012**, 90, 173–176. DOI: 10.1016/j.saa.2012.01.029.
9. Purohit, M. C., Singh, M., Kumar, G., Singh, N. IJPSR, **2020**, *11*, 4524–4529. DOI: 10.13040/IJPSR.0975-8232.11(9).4524–29.
10. Paul, M., Londhe, V. Y. *Applied Organometallic Chemistry*, **2019**, *33*, e4624. DOI: 10.1002/aoc.4624.
11. Aadil, K. R., Pandey, N., Mussatto, S. I., Jha, H. *Journal of Environmental Chemical Engineering*, **2019**, 7, 103296. DOI: 10.1016/j.jece.2019.103296.
12. My-Thao Nguyen, T., Nguyen, T. A.-T., Pham, N. T.-V., Ly, Q.-V., Tran, T. T.-Q., Thach, T.-D., Nguyen, C.-L., Banh, K.-S., Le, V.-D., Nguyen, L.-P., Nguyen, D.-T., Dang, C.-H., Nguyen, T.-D. *Arabian Journal of Chemistry*, **2021**, *14*, 103096. DOI: 10.1016/j.arabjc.2021.103096.
13. Rajeshkumar, S., Menon, S., Kumar S. V., Tambuwala, M. M., Bakshi, H. A., Mehta, M., Satija, S., Gupta, G., Chellappan, D. K., Thangavelu, L., Dua, K. *Journal of Photochemistry and Photobiology, B: Biology*, **2019**, *197*, 111531. DOI: 10.1016/j.jphotobiol.2019.111531.
14. Hashemi, Z., Mohammadyan, M., Naderi, S., Fakhar, M., Biparva, P., Akhtari, J., Ebrahimzadeh, M. A. *Materials Today, Communications*,

**2021**, *27*, 102264. DOI: 10.1016/j.mtcomm.2021.102264.

15. Baruah, K., Haque, M., Langbang, L., Das, S., Aguan, K., Roy, A. S. *Journal of Molecular Liquids*, **2021**, *337*, 116422. DOI: 10.1016/j.molliq.2021.116422.
16. Ma, Z., Liu, J., Liu, Y., Zheng, X., Tang, K. *International Journal of Biological Macromolecules*, **2021**, *166*, 567–577. DOI: 10.1016/j.ijbiomac.2020.10.214.
17. Torres, L. A. Z., Woiciechowski, A. L., Tanobe, V. O. A., Filho, A. Z., Freitas, R. A., Noseda, M. D., Szameitat, E. S., Faulds, C., Coutinho, P., Bertrand, E., Soccol, C. R. *International Journal of Biological Macromolecules*, **2021**, *167*, 1499–1507. DOI: 10.1016/j.ijbiomac.2020.11.104.
18. Ghosal, K., Ghosh, S., Ghosh, D., Sarkar, K. *International Journal of Biological Macromolecules*, **2020**, *162*, 1605–1615. DOI: 10.1016/j.ijbiomac.2020.07.315.
19. Jahana, I., Erci, F., Isildak, I. *Journal of Drug Delivery Science and Technology*, **2021**, *61*, 102172. DOI: 10.1016/j.jddst.2020.102172.
20. Ceylan, R., Demirbas, A., Ocsoy, I., Aktumsek, A. *Sustainable Chemistry and Pharmacy*, **2021**, *21*, 100426. DOI: 10.1016/j.scp.2021.100426.
21. Abdoli, M., Arkan, E., Shekarbeygi, Z., Khaledian, S. *Inorganic Chemistry Communications*, **2021**, *129*, 108649. DOI: 10.1016/j.inoche.2021.108649.
22. Ardakani, L. S., Alimardani, V., Tamaddon, A. M., Amani, A. M., Taghizadeh, S. *Heliyon*, **2021**, *7*, e06159. DOI: 10.1016/j.heliyon.2021.e06159.
23. Olfati, A., Kahrizi, D., Balaky, S. T. J., Sharifi, R., Tahir, M. B., Darvishi, E. *Inorganic Chemistry Communications*, **2021**, 125, 108439. DOI: 10.1016/j.inoche.2020.108439.
24. Muniyappan, N., Pandeeswaran, M., Amalraj, A. *Environmental Chemistry and Ecotoxicology*, **2021**, *3*, 117–124. DOI: 10.1016/j.enceco.2021.01.002.
25. Veeraraghavan, V. P., Periadurai, N. D., Karunakaran, T., Hussain, S., Surapaneni, K. M., Jiao, X. *Saudi Journal of Biological Sciences*, **2021**, *28*, 3633–3640. DOI: 10.1016/j.sjbs.2021.05.007.
26. Naaz, R., Siddiqui, V. U., Qadir, S. U., Siddiqi, W. A. *Materials Today: Proceedings*, **2021**, *46*, 2352–2358. DOI: 10.1016/j.matpr.2021.05.062.
27. Anju, T.R., Parvathy, S., Veettil, M. V., Rosemary, J., Ansalna, T. H., Shahzabanu, M. M., Devik, S. *Materials Today: Proceedings*, **2021**, *43*,

3956–3960. DOI: 10.1016/j.matpr.2021.02.665.

28. Hassan, K. T., Ibraheem, I. J., Hassan, O. M., Obaid, A. S., Ali, H. H., Salih, T. A., Kadhim, M. S. *Journal of Environmental Chemical Engineering*, **2021**, *9*, 105359. DOI: 10.1016/j.jece.2021.105359.
29. Singh, J., Tripathi, J., Sharma, M., Nagar, S., Sharma, A. *Materials Today: Proceedings*, **2021**, *46*, 2294–2297. DOI: 10.1016/j.matpr.2021.04.086.
30. Abdelghaffar, F., Mahmoud, M. G., Asker, M. S., Mohamed, S. S. *Journal of Industrial and Engineering Chemistry*, **2021**, *99*, 224–234. DOI: 10.1016/j.jiec.2021.04.030.
31. AsifAsghar, M., Zahir, E., Shahid, S. M., Khan, M. N., Asghar, M. A., Iqbal, J., Walker, G. *LWT*, **2018**, *90*, 98–107. DOI: 10.1016/j.lwt.2017.12.009
32. Jebril, S., Jenana, R. K. B., Dridi, C. *Materials Chemistry and Physics*, **2020**, *248*, 122898. DOI: 10.1016/j.matchemphys.2020.122898.
33. Ajaz, S., Ahmed, T., Shahid, M., Noman, M., Shah, A. A., Mehmood, M. A., Abbas, A., Cheema, A. I., Iqbal, M. Z., Li, B. *Enzyme and Microbial Technology*, **2021**, *144*, 109745. DOI: 10.1016/j.enzmictec.2021.109745.
34. Shankar, A., Kumar, V., Kaushik, N. K., Kumar, A., Malik, V., Singh, D., Singh, B. *Current Research in Green and Sustainable Chemistry*, **2020**, *3*, 100029. DOI:10.1016/j.crgsc.2020.100029.
35. Chen, J., Li, Y., Fang, G., Cao, Z., Shang, Y., Alfarraj, S., Alharbi, S. A., Li, J., Yang, S., Duan, X. *Arabian Journal of Chemistry*, **2021**, 14, 103000. DOI: 10.1016/j.arabjc.2021.103000.
36. Kambale, E. K., Nkanga, C. I., Mutonkole, B.-P. I., Bapolisi, A. M., Tassa, D. O., Liesse, J.-M. I., Krause, R. W. M., Memvanga, P. B. *Heliyon*, **2020**, *6*, e04493. DOI: 10.1016/j.heliyon.2020.e04493.
37. Alsubki, R., Tabassum, H., Abudawood, M., Rabaan, A. A., Alsobaie, S. F., Ansar, S. *Saudi Journal of Biological Sciences*, **2021**, *28*, 2102–2108. DOI: 10.1016/j.sjbs.2020.12.055.
38. Tailor, G., Yadav, B. L., Chaudhary, J., Joshi, M., Suvalka, C. *Biochemistry and Biophysics Reports*, **2020**, *24*, 100848. DOI: 10.1016/j.bbrep.2020.100848.
39. Rodríguez-Felix, F., López-Cota, A. G., Moreno-Vásquez, M. J., Graciano-Verdugo, A. Z., Quintero-Reyes, I. E., Del-Toro-Sánchez, C. L., Tapia-Hernández, J. A. *Heliyon*, **2021**, *7*, e06923. DOI: 10.1016/j.heliyon.2021.e06923.
40. Sharma, V., Kaushik, S., Pandit, P., Dhull, D., Yadav, J. P., Kaushik, S. *Applied Microbiology and Biotechnology*, **2019**, *103*, 881–891. DOI:

10.1007/s00253-018-9488-1.

41. Haggag, E. G., Elshamy, A. M., Rabeh, M. A., Gabr, N. M., Salem, M., Youssif, K. A., Samir, A., Muhsinah, A. B., Alsayari, A., Abdelmohsen, U. R. *International Journal of Nanomedicine*, **2019**, *14*, 6217–6229. DOI: 10.2147/IJN.S214171.
42. Mehmood, Y., Farooq, U., Yousaf, H., Riaz, H., Mahmood, R. K., Nawaz, A., Abid, Z., Gondal, M., Malik, N. S., Barkat, K., Khalid, I. *Pakistan Journal of Pharmaceutical Sciences*, **2020**, *33*, 839–845.
43. Jain, D., Kothari, S. L. *Journal of Mycology and Plant Pathology*, **2014**, *44*, 21–24.
44. Mali, S. C., Dhaka, A., Githala, C. K., Trivedi, R. *Biotechnology Reports*, **2020**, *27*, e00518. DOI: 10.1016/j.btre.2020.e00518.
45. Ali, M., Haroon, U., Khizar, M., Chaudhary, H. J., Munis, M. F. H. *Current Plant Biology*, **2020**, *23*, 100157. DOI: 10.1016/j.cpb.2020.100157.
46. Umamaheswari, A., Prabu, S. L., John, S. A., Puratchikody, A. *Biotechnology Reports*, **2021**, *29*, e00595. DOI: 10.1016/j.btre.2021.e00595.
47. Manasa, D.J., Chandrashekara, K. R., Madhu Kumar, D. J., Niranjana, M., Navada, K. M. *Arabian Journal of Chemistry*, **2021**, *14*, 103184. DOI: 10.1016/j.arabjc.2021.103184.
48. Ramesh, P., Saravanan, K., Manogar, P., Johnson, J., Vinoth, E., Mayakannan, M. *Sensing and Bio-Sensing Research*, **2021**, *31*, 100399. DOI: 10.1016/j.sbsr.2021.100399.
49. Das, P., Ghosh, S., Ghosh, R., Dam, S., Baskey(Sen), M. *Journal of Photochemistry and Photobiology, B: Biology*, **2018**, *189*, 66–73. DOI: 10.1016/j.jphotobiol.2018.09.023.
50. Resmi, R., Yoonus, J., Beena, B. *Materials Today: Proceedings*, **2021**, *46*, 3062–3068. DOI: 10.1016/j.matpr.2021.02.498.
51. Sharma, S., Kumar, K., Thakur, N., Chauhan, S.,. Chauhan, S. M. *Journal of Environmental Chemical Engineering*, **2021**, *9*, 105395. DOI: 10.1016/j.jece.2021.105395.
52. Pandiyan, N., Murugesan, B., Arumugam, M., Chinnaalagu, D., Samayanan, S., Mahalingam, S. *Advanced Powder Technology*, **2021**, *32*, 2213–2225 DOI: 10.1016/j.apt.2021.04.030.
53. Hussein, B. Y., Mohammed, A. M. *Materials Today:Proceedings*, **2021**, *42*, A18–A26. DOI: 10.1016/j.matpr.2021.03.729.
54. Thakur, S., Shandilya, M., Guleria, G. *Journal of Environmental Chemical*

*Engineering*, **2021**, *9*, 104882. DOI: 10.1016/j.jece.2020.104882.

55. Chennimalai, M., Vijayalakshmi, V., Senthil, T. S., Sivakumar, N. *Materials Today: Proceedings*, **2021**. DOI: 10.1016/j.matpr.2021.03.409.
56. Skheel, A. Z., Jaduaa, M. H., Abd, A. N. *Materials Today: Proceedings*, **2021**, *45*, 5793–5799. DOI: 10.1016/j.matpr.2021.03.168.
57. Arumugam, M., Manikandan, D. B., Dhandapani, E., Sridhar, A., Balakrishnan, K., Markandan, M., Ramasamy, T. *Environmental Technology and Innovation*, **2021**, *23*, 101653. DOI: 10.1016/j.eti.2021.101653.
58. Shreema, K., Mathammal, R., Kalaiselvi, V., Vijayakumar, S., Selvakumar, K., Senthil, K. *Materials Today: Proceedings*, **2021**. DOI: 10.1016/j.matpr.2021.04.627.
59. Jobie, F. N., Ranjbar, M., Moghaddam, A. H., Kiani, M. *Advanced Powder Technology*, **2021**, *32*, 2043–2052. DOI: 10.1016/j.apt.2021.04.014.
60. Chai, H.-Y., Lam, S.-M., Sin, J.-C. *Materials Letters*, **2019**, *242*, 103–106.
61. Madhumitha, G., Fowsiya, J., Gupta, N., Kumar, A., Singh, M. *Journal of Physics and Chemistry of Solids*, **2019**, *127*, 43–51. DOI: 10.1016/j.jpcs.2018.12.005.
62. Irshad, M. A., Nawaz, R., Rehman, M. Z., Imrand, M., Ahmade, J., Ahmad, S., Inam, A., Razzaq, A., Rizwan, M., Ali, S. *Chemosphere*, **2020**, *258*, 127352. DOI: 10.1016/j.chemosphere.2020.127352.
63. Ahmed, T., Ren, H., Noman, M., Shahid, M., Liu, M., Ali, M. A., Zhang, J., Tian, Y., Qi, X., Li, B. *Nano Impact*, **2021**, *21*, 100281. DOI: 10.1016/j.impact.2020.100281.
64. Pillai, A. M., Sivasankarapillai, V. S., Rahdar, A., Joseph, J., Sadeghfar, F., Anuf A, R., Rajesh, K., Kyzas, G. Z. *Journal of Molecular Structure*, **2020**, *1211*, 128107. DOI: 10.1016/j.molstruc.2020.128107.
65. Jamdagni, P., Khatri, P., Rana, J. S. *Journal of King Saud University – Science*, **2018**, *30*, 168–175. DOI:10.1016/j.jksus.2016.10.002.

# Chapter 9

# Green Nanomaterials in Energy Applications and Sensor Implementations

**Kaushik Pal**
*University Centre for Research and Development (UCRD),*
*Department of Physics, Chandigarh University,*
*Mohali, Gharuan, Punjab, India*
kaushik.e12567@cumail.in

The novel ideas of green nanotechnology have attracted the research community owing to its enormous industrial applications. The extraction of naturally occurring resources from environmental precursors has resulted in the manufacture of cutting-edge nanomaterials in an easy and economical way. This chapter examines the various methodologies for the green synthesis of zinc oxide (ZnO) using *Ihlamur* leaf extract, $TiO_2$ using *Bixa orellana* seed extract and Ag nanoparticles using *neem leaf* extract and its promising applications in the field of energy and in sensors. The greater pore volume and long range surface area observed in the green nanoparticles are attributed to the green route of the synthesis methods. This effective biosynthesis substantially eliminates the

*Green Nanoarchitectonics*
Edited by Kaushik Pal

ISBN 978-981-4968-50-8 (Hardcover), 978-1-003-31860-6 (eBook)
www.jennystanford.com

usage of toxic reagents and employs natural resources to create nanomaterials with potential applications, implying a cost-effective and environmentally friendly solution. A detailed investigation of the utilization of ZnO and $TiO_2$ nanoparticles as an inevitable ingredient in dye-sensitized solar cells and their efficiency has been explored through various characterizations. Sensing applications of green silver nanoparticles for detecting ammonia dissolved in water and pesticide detection are also well-explored in this chapter.

## 9.1 Introduction

Nanotechnology, which encompasses the development and utilization of materials with at least one dimension limited to between 1 and 100 nm, has sparked much interest in both research and industry. Nanomaterials with different physical and chemical properties can have their structural features altered to achieve unique functions. The characteristics of nanomaterials differ significantly from those of their bulk counterparts and use significantly less raw materials, cost, and energy to attain the same goal in comparison to bulk materials [1]. Nanoparticles display novel and enhanced properties as a result of their nanoscale dimension, molecule dispersion, and shape. Nanotechnology's importance in today's world has been highlighted by the synthesis of nanoparticles of various sizes, shapes, and chemical compositions. Chemical vapor deposition, laser ablation, etching, supercritical fluid, sol-gel process, pyrolysis, and electro deposition are some of the conventional chemical and physical methods for synthesizing nanoparticles. The chemical reduction of numerous metals and compounds is used to create nanoparticles. Although the nanoparticles produced have a wide range of applications, these synthesis processes have a number of disadvantages, including excessive energy consumption, the use of environmentally dangerous chemicals, high cost, and the formation of poisonous byproducts [1, 2]. Researchers suggest that nanoparticles can cause harm at the cellular, subcellular, and molecular levels and impose health risks and environmental impacts when directly and indirectly exposed to nanoparticles [3, 4]. Chemical precursors penetrating nanoparticle surfaces make it challenging to use chemically generated nanoparticles in biomedicine. These factors prompted the development of novel synthesis methods of nanoparticles that have a lower environmental impact, as they encourage the manufacture of nanomaterials to avoid the usage of harmful chemicals and instead rely on natural resources, nontoxic solvents, and energy-efficient synthesis methods [5, 6]. Green nanotechnology is significant because it minimizes nanoparticle toxicity and expands the biomedical applications of nanoparticles by using environmentally acceptable and natural sources that act as reducing and stabilizing agents, making biosynthesized nanomaterials more biocompatible [7]. Plant extract-based synthesis of nanomaterials has sparked much attention since it is faster, easier, and less expensive than utilizing microorganisms, and it allows for the creation of green nanomaterials with superior size and morphology [8–10]. Due to the great diversity of microorganisms that react to metal ions in diverse ways for the creation of metal nanoparticles (MNPs), microbes are also crucial for the green synthesis of MNPs. When compared to traditional chemical approaches, the green synthesis of MNPs employing microorganisms has various advantages, as it provides a method for the synthesis of MNPs that is fast, cost-effective, clean, nontoxic, and good for the environment [11, 12]. The existence of many elements in microorganisms, such as enzymes, proteins, and other biological molecules, plays an essential role in the reduction of MNPs. Furthermore, proteins released by microorganisms can serve as stabilizing agents, ensuring colloidal stability and inhibiting MNP agglomeration [13].

## 9.2 Synthesis of Nanomaterials Using Green Chemistry Approach

Green synthesis of nanomaterials is important for future prospects, as this field of nanoscience should lead to the production of safe, environmentally acceptable nanoparticles [14]. Figure 9.1 demonstrates various methods employed for the development of nanoparticles. The two distinct methods for the development of nanoparticles are "top-down" and "bottom-up." Using various procedures, such as grinding, milling, sputtering, thermal/laser ablation, and others, a suitable bulk material is broken down into smaller fine particles in the "top-down" method. Nanoparticles are generated by chemical and biological techniques by the self-assembly of atoms to new nuclei, which grow into nanosized particles in the bottom-to-top approach. Chemical reduction, electrochemical procedures, and sonodecomposition are examples of "bottom-up" approaches. Solvents and reducing agents used to reduce nanoparticles have a significant impact on the morphology of integrated particles [15]. It is necessary to create and develop greener synthesis of nanoparticles to comprehend nanotechnologies that cause little harm to human health or the environment, as well as to cultivate technologies that can be used to protect the environment [16]. Green nanomaterial synthesis is a bottom-up process in which the primary reaction is oxidation or reduction [17]. Plant extracts exhibit excellent oxidizing or reducing properties, making them ideal for converting metal compounds into their nanoparticles. The biocompatibility of green nanomaterials is enhanced by the coating of biological molecules on the surface of nanoparticles [18]. The synthesis of nanoparticles employing several types of microorganisms has been proven to be safer and more cost efficient. Green-synthesized nanoparticles are currently preferred over traditionally supplied nanoparticles due to their excellent qualities.

### 9.2.1 Green Synthesis of Nanomaterials Using Microbes

Biological resources offer a flexible, cost-effective, and environmentally acceptable way to make MNPs with remarkable physical, chemical, and biological properties [19]. Microorganisms such as bacteria, fungi, and yeast can also be used to make green nanoparticles by exploiting their defense mechanisms. Nanoparticle production is caused by the resistance of bacterial cells to reactive ions in the environment. Because bacterial cells are vulnerable to high ion concentrations, reactive ions are transformed to stable atoms to prevent cell death. This feature of bacteria is being used for green nanoparticle production [20]. Metals are hydrolyzed by enzymes released by unicellular and multicellular microorganisms, resulting in enzymatic metal ion reduction and the formation of metallic nanoparticles with desired morphology [21, 22]. Although nanoparticle synthesis using microbes is a slow and expensive process, it is environmentally friendly and compatible and can be used in a variety of ways. In the synthesis of MNPs, cofactors such as nicotinamide adenine dinucleotide (NADH) and the reduced form of nicotinamide dinucleotide phosphate (NADPH)-dependent enzymes operate as reducing agents by transferring electrons from NADH by NADH-reliant enzymes, which act as electron carriers [6].

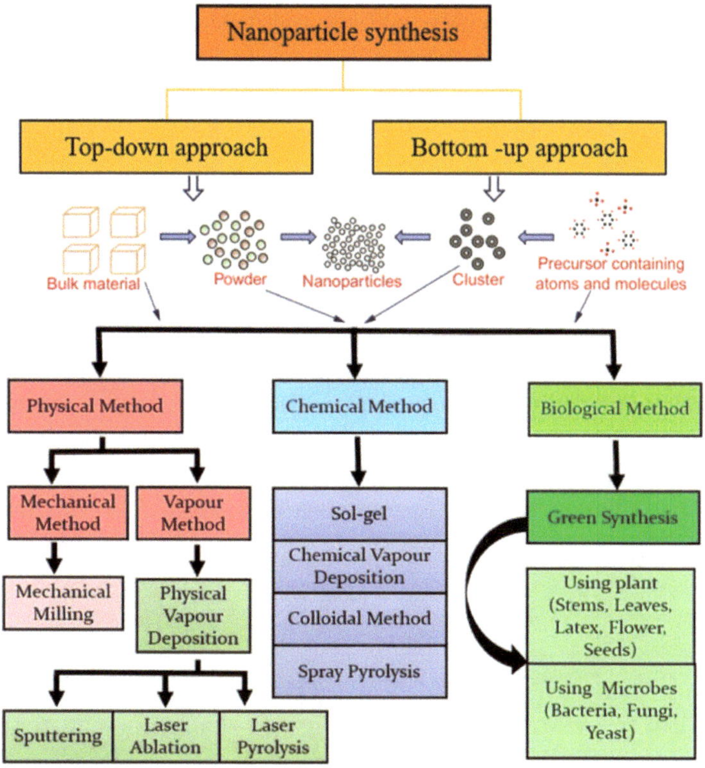


**Figure 9.1** Different approaches for the synthesis of nanoparticles. Reprinted with permission from Ref. 23, Copyright 2018, Elsevier.

### 9.2.2 Plant Extract-Based Biosynthesis of Nanomaterials

Plant extracts from diverse sections, such as the leaf, root, latex, seed, and stem, act as stabilizing or reducing agents in the production of nanoparticles. These nanoparticles have the potential to be employed in cancer treatment and antioxidant therapy [25]. Green tea leaf extract [26], *Terminalia chebula* fruit extract [27], oolong and black tea leaf extract [28], banana peel extract with *Colocasia esculenta* leaf extract [29], sorghum bran extract [29], eucalyptus leaf extract [29], and *Tridax procumbens* leaf extract [30] have all been used to make nanoparticles. According to studies, plant extracts and their biomolecules exhibit remarkable reducing properties that aid in the reduction of nanoparticles [31]. Because plant extracts have a high reduction capacity for metal ions, it has been discovered that stable MNPs can be easily generated. Plant-derived biologically active chemicals such as phenolic acids, amino acids, polyphenols, saponins, proteins, flavones, sugars, terpenoids, and so on, serve as a capping and stabilizing agent in the manufacture of MNPs [8]. The features of the plant extract, such as its content, metal salt concentration, duration of reaction, reaction solution pH, and temperature, determine the quality, size, and shape of the synthesized nanoparticles. Figure 9.2 shows the preparation of gold nanoparticles using tamarind juice extract, where 3 ml of Tamarindus fruit essence was mixed with 2 ml of tetrachloroauric acid precursor. Meanwhile, the solution turned dark brown, indicating the formation of gold nanoparticles with an average diameter of 500–520 nm.

**Figure 9.2** Using tamarind juice extract to demonstrate the various procedures required in the biosynthesis of gold nanoparticles. Reprinted with permission from Ref. 32, Copyright@ 2020, Elsevier.

## 9.3 Potential Trends of Green Nanomaterials in Energy Applications

A reliable and abundant supply of energy is needed for daily life, and it must be delivered with minimal negative environmental consequences. Environmental pollution has been proven to have serious consequences for the climate and the global economy to date. As a result, it is vital to substitute renewable and ecologically friendly energy sources for dangerous fossil fuels. Dye-sensitized solar cells (DSSCs) have emerged as the most reliable renewable energy resource in recent years due to their high incident solar light-to-electricity conversion efficiency.

### 9.3.1 Dye-Sensitized Solar Cells Based on Green-Synthesized Zinc Oxide Nanoparticles

Because of its unique optical, electrical, and semiconductor capabilities, as well as its broad bandgap and massive exciton binding energy, zinc oxide (ZnO) nanoparticles have potential uses in energy applications [33, 34]. Traditional chemical methods for synthesizing ZnO nanoparticles require the use of expensive apparatuses, dangerous chemicals, and high temperatures and may result in contaminants. The green synthesis of ZnO nanoparticles is easy, ecologically safe, nonhazardous, uncomplicated and economical, and it has a number of advantages over other nanoparticle preparation techniques. This section shows how *Tilia tomentosa* (Ihlamur) leaf extract is used to make ZnO nanomaterials and how it is used in solar cells.

#### 9.3.1.1 Tilia tomentosa (Ihlamur) leaf extract for the biosynthesis of ZnO nanoparticles

Figure 9.3 shows the various steps involved in the green synthesis of ZnO. Here, 10 g of powdered dried leaves of *Tilia tomentosa* were combined with 200 ml of distilled water and heated for 15 minutes at 80°C until a bright yellow tinted solution remained. After cooling to room temperature and filtering, a clear yellow-colored plant extract solution was obtained, which served as a reducing and capping agent. Twenty milliliters of plant extract was mixed with 0.1 M zinc nitrate hexahydrate solution at 80°C and stirred constantly for approximately 30 minutes on a magnetic stirrer. NaOH (1 M) was added dropwise, and after 30 minutes of stirring, the yellow color solution transformed into a pale yellow color precipitate. Total precipitation was achieved by continuing to stir at 80°C for another 1 or 2 minutes. The solution was then allowed to cool to room temperature before being centrifuged at 7500 rpm for 10 minutes. The resultant solution was extracted and centrifuged once more after being rinsed with water. After three centrifugation cycles, the white ZnO nanoparticles were collected and desiccated overnight at 70°C in an oven. The ZnO nanoparticles were subjected to a temperature of 400°C for 15 minutes to eliminate any evaporable impurities [35].

#### 9.3.1.2 DSSC implementation using green-synthesized ZnO nanoparticles

ZnO nanoparticles were combined with a small amount of Triton-X 100, acetylacetone, and distilled water to make a paste. Using the doctor blade process, this paste was coated onto fluorine-doped tin oxide (FTO) with a surface area of 0.25 $cm^2$. A Doctor blade process

**Figure 9.3** Illustrating the biosynthesis of ZnO nanoparticles using *Ihlamur leaf extract.*

was used to squeeze the paste onto the FTO surface, forming a ZnO film coating. To create a sandwich-type DSSC, as shown in Figure 9.4, the dye-absorbed ZnO film was clipped with a platinum-coated counter electrode, and sealing tape was employed to avoid short circuiting. The redox electrolyte was introduced into the device through the space in the counter electrode [35].

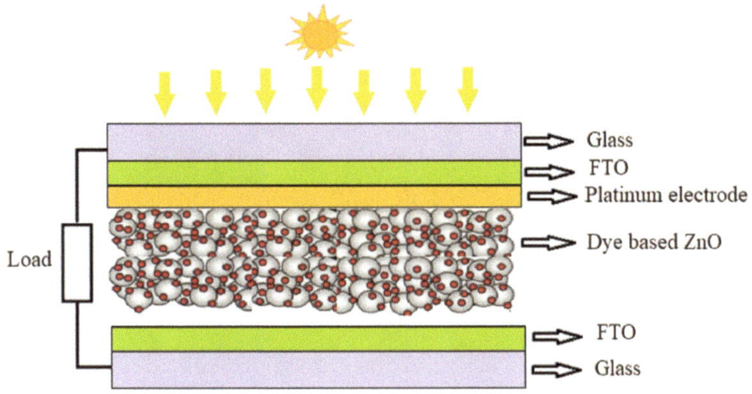


**Figure 9.4** Showing the structure of ZnO-based DSSCs.

### 9.3.1.3 Spectroscopic and microscopic analysis of ZnO nanoparticles and DSSCs

The scanning electron microscopic (SEM) image of green-synthesized ZnO nanoparticles generated by Ihlamur leaves is shown in Fig. 9.5a. The sphere-shaped nature of the produced nanoparticles with identical diameters is depicted in the SEM microstructure. The ZnO nanoparticles possess an average diameter of approximately 80 nm. Because of their small size, ZnO nanoparticles are naturally homogenous with negligible aggregation. The energy dispersive spectroscopy (EDS) image of ZnO nanoparticles produced using Ihlamur plant extract, as shown in Fig. 9.5b, was used to determine the elemental composition of the ZnO nanoparticles. The atomic percentages of zinc and oxygen were calculated experimentally to be 50% each. The X-ray diffraction (XRD) pattern of the produced ZnO nanoparticles depicted in Fig. 9.5c shows that the diffraction peaks at $2\theta$ values of 31.71°, 34.37°, and 36.22° correspond to the (100), (002), and (101) planes, respectively, confirming the hexagonal structure of ZnO. The UV-visible spectrum displayed in Fig. 9.5d was used to determine the optical band gap of ZnO nanoparticles.

Because of the surface plasmon absorption of ZnO nanoparticles, the spectrum reveals a large absorption peak at 350 nm. When electromagnetic radiation strikes on the free conduction band electrons, they collectively oscillate, resulting in surface plasmon absorption [36, 37]. The calculated bandgap energy of the ZnO nanoparticles was 3.55 eV.

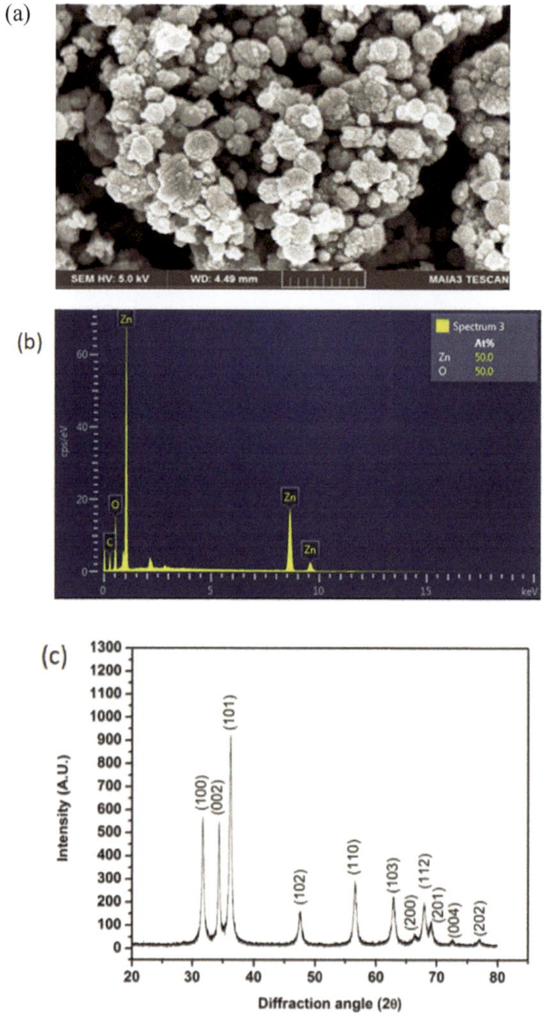


**Figure 9.5** *(continued)*

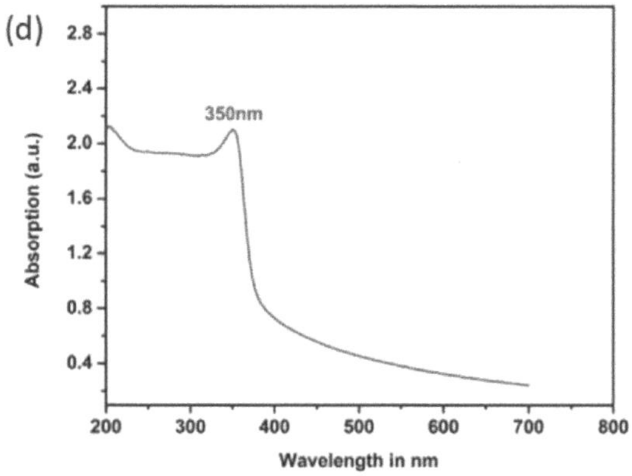


**Figure 9.5** Spectroscopic and microscopic characterization. (a) The scanning electron microscopic (SEM) image of green-synthesized ZnO nanoparticles indicating the sphere-shaped nature of the nanoparticles with identical diameters, (b) the energy dispersive spectroscopy (EDS) image of ZnO nanoparticles used to determine the elemental composition of the ZnO nanoparticles, (c) the X-ray diffraction (XRD) pattern of the produced ZnO nanoparticles, and (d) the UV–visible spectrum of the green ZnO nanoparticle showing a large absorption peak at 350 nm. Reprinted with permission from Ref. 35, Copyright@ 2020, Elsevier.

#### 9.3.1.4 Electrical characterization of DSSCs constructed using green-synthesized ZnO

The current density–voltage (*J–V*) curve of the ZnO-based DSSC at various light intensities is illustrated in Fig. 9.6. DSSC parameters such as open-circuit voltage ($V_{OC}$), short circuit current density ($J_{SC}$), fill factor (*FF*), and efficiency (*h*) can be determined from the current density–voltage curves, and these calculated values are shown in Table 9.1.

**Table 9.1** Displaying the photovoltaic parameters of ZnO-based DSSCs

| Light intensity ($mW/cm^2$) | $J_{SC}$ ($mA/cm^2$) | $V_{OC}$ (V) | η (%) | *FF* (%) |
|---|---|---|---|---|
| 20 | 0.67 | 0.49 | 1.01 | 61.3 |
| 40 | 1.90 | 0.59 | 1.11 | 35.9 |
| 60 | 3.37 | 0.61 | 1.41 | 41.3 |
| 80 | 5.10 | 0.62 | 1.65 | 41.6 |
| 100 | 6.26 | 0.65 | 1.97 | 48.5 |

*Source*: Reprinted with permission from Ref. 35, Copyright@ 2020, Elsevier.

The fill factor is calculated using the equation:

$$\text{Fill factor } (FF) = \frac{J_{max} \times V_{max}}{J_{SC} \times V_{OC}},$$

where $J_{max}$ and $V_{max}$ are the current density and the voltage for the maximum power, respectively, $J_{SC}$ is the short current density at 0 V and $V_{OC}$ is the open circuit voltage.

The power conversion efficiency ($\eta$) is calculated using the equation:

$$\eta = \frac{FF \times V_{OC} \times J_{SC}}{S \times P_{in}},$$

where $S$ is the area of the solar cell.

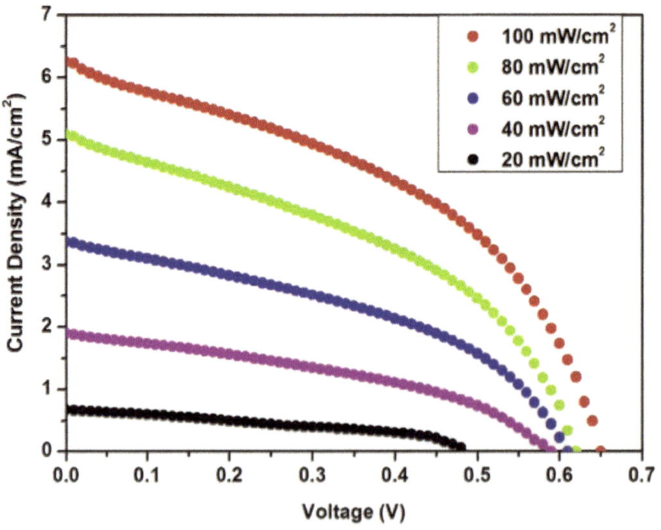


**Figure 9.6** Demonstrating the current density–voltage (J–V) curve of the ZnO-based DSSC at various light intensities. Reprinted with permission from Ref. 35, Copyright@ 2020, Elsevier.

When compared to previous similar studies, the percentage efficiency of the green ZnO nanoparticle-based DSSC was found to be fairly good. Enhanced absorption owing to more dye molecules being adsorbed on the ZnO surface is one of the factors responsible for the high efficiency. The utilization of ZnO nanoparticles in solar cells has a bright future since the biosynthesis of ZnO nanoparticles using plant extract is simple and easy [35].

### 9.3.2 The Use of *Bixa orellana* Seed Extract in the Green Production of $TiO_2$ Nanoparticles and Their Application in Solar Cells

Due to its unique qualities, such as well-matched band alignment with dyes, recyclability, great charge transfer capability, and high refractive index, $TiO_2$ has been demonstrated to be one of the most promising photoanode materials in several research studies [38]. To enhance the overall performance of the solar cell, different capping agents are utilized to produce $TiO_2$ that avoids agglomeration and reduces the average particle size. Biocapping agents allow for an environmentally friendly, easy synthesis of nanoparticles in an economical way [39]. This section covers how *B. orellana* seed extract is used to make $TiO_2$ nanomaterials and how it is used in solar cells.

#### 9.3.2.1 $TiO_2$ synthesis utilizing B. orellana seed extract in a green route

Figure 9.7 shows the step-by-step procedure for the fabrication of $TiO_2$ nanoparticles using *B. orellana* extract. The dried *B. orellana* seed was crushed into a fine powder, and 2 g of this powder was soaked in 100 ml of ethanol for 1 day in the dark. The extract was filtered and stored for the preparation of nanoparticles. Ten milliliters of titanium butoxide was added to 80 ml of isopropanol, and the mixture was refluxed for 4 hours at 70°C to obtain a transparent solution. This solution was then cooled to room temperature, and 5 ml of glacial acetic acid was added. The obtained solution was refluxed for 2 hours at 70°C for 2 hours and then cooled to room temperature before adding 5 ml of *B. orellana* seed extract. After that, the mixture was refluxed for 2 hours at 70°C, cooled to room temperature and then, under vigorous stirring, dropped into 15 ml deionized water drop by drop. The blended solution was mixed continuously for 2 hours and then stored at room temperature for 24 hours. Centrifugation was used to recover the precipitates, which were then extensively cleaned with ethanol and deionized water before being dried for 24 hours in an oven at 60°C. The powder was then calcined for 1 hour at 450°C [40].

### 9.3.2.2 Fabrication of a DSSC based on green-synthesized $TiO_2$

Two hundred milligrams of $TiO_2$ powder, 1 drop of Triton X100, and 3 drops of acetic acid were combined to make $TiO_2$ paste. The mixture was vigorously mixed for 10 minutes, followed by the addition of 2 ml of ethanol slowly while mixing proceeded for another 10 minutes. After that, the paste was placed on a clean FTO substrate. The paste was then applied using a doctor-blade method to a clean FTO-coated clear conducting glass to produce a $TiO_2$ film with a thickness of 10 μm and an area of 0.25 $cm^2$. The $TiO_2$ film was allowed to settle at room temperature for one day before being annealed in air at 450°C for half an hour and then cooled to room temperature. Deposition of Pt catalyst on a conductive glass plate and annealing at a temperature of 400°C for 30 minutes results in the formation of a platinum counter electrode [40].

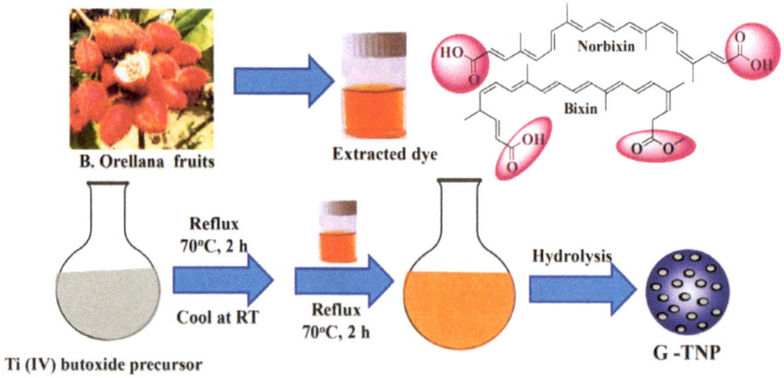


**Figure 9.7** This study demonstrates the fabrication of $TiO_2$ nanoparticles using *B. orellana seed extract.* Reprinted with permission from Ref. 40, Copyright@ 2020, Elsevier.

$TiO_2$ sheets were then coated with N719 dye and assembled into a sandwich structure for DSSC devices. The counter electrode was placed on top of the cell, with its conductive side facing the dye-coated $TiO_2$ sheet, and the cell was sealed on three sides with the spacer, with one side open for electrolyte solution injection. The liquid electrolyte was a 0.2 M lithium iodide and 0.02 M iodine in acetonitrile solution that was introduced into the constructed cells through the open side and pulled into the space between the electrodes by capillary action. Silver paste was used to adhere the copper wires to the electrode, which was then covered with Araldite and sealed [40].

#### 9.3.2.3 Spectroscopic characterization and optical properties of biosynthesized $TiO_2$

Figure 9.8a shows the XRD pattern of $TiO_2$ nanoparticles produced by conventional and green chemistry methods. Sharp peaks for $TiO_2$ nanoparticles made by the green synthesis method at 2θ values of 25.3, 37.8, 48.1, 53.9, 55.1, 62.7, 68.8, 70.3, and 75.1 match the crystal planes (1 0 1), (0 0 4), (2 0 0), (1 0 5), (2 1 1), (2 0 4), (1 1 6), (2 1 5), and (3 0 3), indicating that the biosynthesized $TiO_2$ nanoparticles are extremely crystalline and that the average crystalline size was found to be 9 nm. The functional groups present in the synthesized compounds were confirmed using FTIR spectroscopy, as shown in the Fourier transform infrared radiation spectra (FT-IR) of TNP and G-TNP (Fig. 9.8b). The O–Ti–O lattice vibration of $TiO_2$'s anatase phase is responsible for the peaks in the 400–700 $cm^{-1}$ range [41]. The peak at 1630 $cm^{-1}$ corresponds to a H–O bond bending vibration [42]. The stretching vibration of -OH and weakly bound water molecules is assigned to the broad band detected at 2900–3400 $cm^{-1}$ [43].

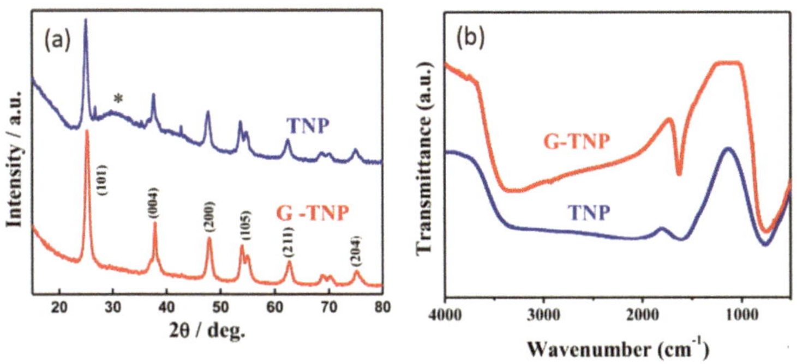


**Figure 9.8** (a) X-ray diffraction (XRD) pattern and (b) Fourier transform infrared spectroscopy (FT-IR) spectra of $TiO_2$ nanoparticles produced by conventional and green chemistry methods. Reprinted with permission from Ref. 40, Copyright@ 2020, Elsevier.

Scanning electron microscopy (SEM) and high resolution transmission electron microscopy (HR-TEM) images, as shown in Fig. 9.9a,b, respectively, revealed that green-synthesized $TiO_2$ nanoparticles had irregular spherical morphologies and rougher surfaces, with an average particle size of 13 + 2 nm. It is thus clear that *B. orellana* has both stabilizing and capping effects on the shape, microstructures, and size of $TiO_2$ nanoparticles produced. All of these investigations show that the green synthesis approach is a highly effective way to manufacture nanosized pure anatase $TiO_2$ nanoparticles with mesoporous structural features and huge charge transfer and dye adsorption capabilities [40].

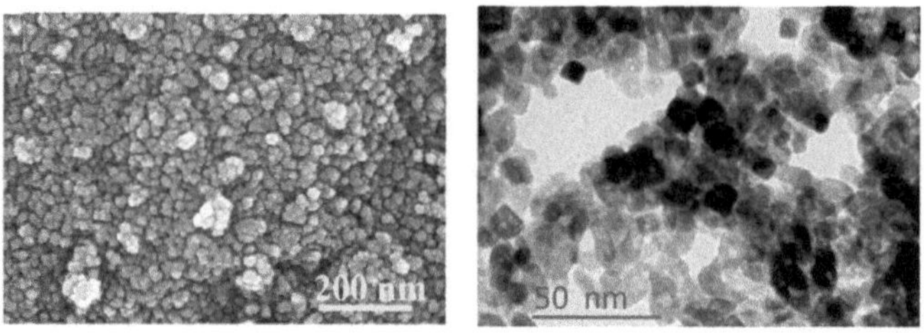


**Figure 9.9** (a) Field emission scanning electron microscopy (FESEM) and (b) HRTEM images depict the irregular spherical morphologies and nonuniform surface of the green-synthesized $TiO_2$ nanoparticles. Reprinted with permission from Ref. 40, Copyright@ 2020, Elsevier.

### 9.3.2.4 Photovoltaic performance of a solar cell based on biosynthesis $TiO_2$

The photocurrent density–voltage (J–V) characteristic curves of the DSSC utilizing $TiO_2$ nanoparticles and green-synthesized $TiO_2$ nanoparticles are shown in Fig. 9.10a, and the corresponding power–voltage curves are shown in Fig. 9.10b. Table 9.2 summarizes the photovoltaic output parameters.

**Table 9.2** A comparison of output parameters of photovoltaic cell and dye loading

| Sample | $J_{SC}$ (mA/cm$^2$) | $V_{OC}$ (mV) | *FF* (%) | η (%) | Amount of dye (mol/cm$^2$) |
|---|---|---|---|---|---|
| TNP | 2.0 | 680 | 73 | 1.03 | $5.12 \times 10^{-8}$ |
| G-TNP | 9.0 | 506 | 65 | 2.97 | $8.56 \times 10^{-8}$ |

*Source*: Reprinted with permission from Ref. 40, Copyright@ 2020, Elsevier.

$J_{sc}$ is the photocurrent density, $V_{oc}$ is the open-circuit voltage, *FF* is the fill factor and is the overall cell efficiency of the DSSC. The $J_{sc}$ value of the biosynthesized $TiO_2$-based solar cell was found to be significantly enhanced from 2 to 9 mA/cm$^2$ when compared to $TiO_2$-based devices, which resulted in a higher cell efficiency of approximately 2.97%. A small reduction in the open circuit voltage from 0.680 to 0.506 V suggested greater electron recombination in green-synthesized $TiO_2$-based solar cells. The increase in efficiency is attributed to various factors, such as its crystalline nature, and it contains only pure anatase, which aids in electron transport. Additionally, the mesoporous structure and large surface area of the biosynthesized $TiO_2$ sheet facilitates easy and effective interfacial electron transport between $TiO_2$, dye, and the electrolyte's redox active species. As a result, it can be stated that using the *B. orellana* dye extract as a capping and stabilizing agent to generate a mesoporous structure and pure anatase phase in $TiO_2$ is an effective method for improving the photovoltaic performance [40].

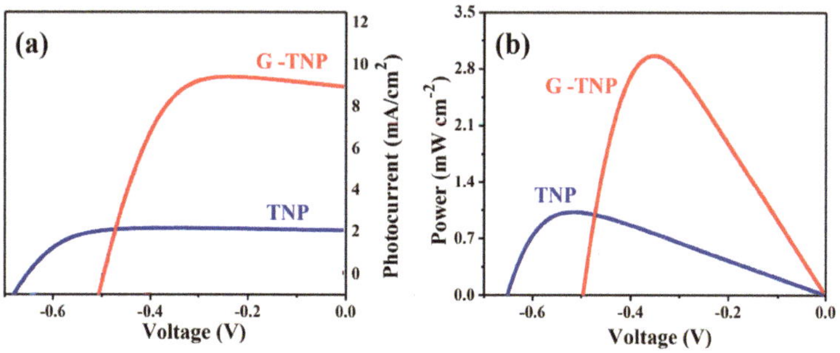


**Figure 9.10** (a) Photocurrent-voltage characteristic curve and (b) power–voltage curve of the green-synthesized $TiO_2$-based solar cell in comparison with the chemically synthesized $TiO_2$-based solar cell. Reprinted with permission from Ref. 40, Copyright@ 2020, Elsevier.

## 9.4 Sensor Implementation and Performance Analysis Using Green-Synthesized Nanomaterials

Recognizing chemical and biological compounds is crucial in biomedical, forensic, and environmental research, and this necessitates advanced equipment as well as fundamental scientific understanding for the development of extremely sensitive and cost-effective sensors. Due to their unique physicochemical features, nanoparticles, particularly noble MNPs (Au, Ag), have attracted much attention in this field and can be used to build revolutionary identification and transduction processes for chemical and biological sensors. Traditional synthesis uses harmful chemicals, which limits their application in real samples. As a result, the scientific community is interested in green nanoparticle synthesis and its applications in the field of sensors. This chapter discusses important applications of green silver nanoparticles (Ag NPs) as sensors for detecting the presence of ammonia and pesticides.

### 9.4.1 The Use of Neem Leaf Extract in the Green Production of Ag NPs and Their Application in Sensing

Strong conductivity, good light absorption, high sensitivity, resolution, antibacterial activity, and chemical stability make Ag NPs as one of the most investigated MNPs [44, 45]. Ag NPs are suitable for biosensing applications because their surface plasmon resonance (SPR) is highly influenced by surface-adsorbed molecules, and the wavelength of SPR may be adjusted by the amount of adsorbed molecules [46]. The increased production cost and toxic effect of physical and chemical approaches to silver nanomaterial synthesis, on the other hand, paved the way for a much more economical, dependable, and environmentally friendly green synthesis method that uses green leaf extract as a reducing and stabilizing agent [44, 47, 48]. Silver ions are reduced to Ag NPs with the help of proteins and enzymes found in the leaf.

#### 9.4.1.1 Silver nanoparticles synthesized using neem leaf extract

The neem leaves were rinsed several times with purified water to eliminate any impurities and then naturally dried after removing the contaminants. After that, 50 ml of distilled water was added to 5 g of chopped leaves in a beaker and heated for 15 minutes at 75°C. To acquire the leaf extract, it was allowed to cool naturally, stirred for 24 hours, and then filtered. Ten milliliters of leaf extract was mixed with silver nitrate solution and rapidly stirred for 10 minutes at room temperature. The colorless solution had now changed to a golden brownish solution. Finally, to obtain Ag NPs, this solution mixture was exposed to sunlight for 15 minutes. The color change from golden brown to dark brown indicates the production of Ag NPs. Figure 9.11 shows all the stages involved in the biosynthesis of Ag NPs from neem leaves [49].

**Figure 9.11** Demonstration of various steps associated with the production of Ag NPs using neem leaves.

### 9.4.1.2 Characterization of Ag NPs synthesized using neem leaves

The shape and size of the Ag NPs and shows that the polyphenol biomolecule layer surrounds the evenly distributed Ag NPs are capped and dispersed by the biomolecules present in the leaf extract. The EDS investigation also confirmed the presence of Ag NPs, as indicated by the long peak (69.4%) along with the presence of other elements, such as oxygen (10.18%), carbon (14.37%) and nitrogen (6.11%).

The existence of biomolecules attached to the surface of Ag NPs is responsible for the presence of oxygen, carbon, and nitrogen peaks [50]. The FTIR spectrum shows absorption bands at 3351, 2948, 1637, 1439, 1211, and 1058 $cm^{-1}$ corresponding to O–H stretching,

C–H stretching, N–H bending, C–H bending, C=O stretching, and C–N stretching, with various functional groups, such as amines, carboxylic acids, aldehydes, ketones, alcohols, and aliphatic amines, adsorbed on the surface of nanoparticles. The SPR bands centered at 442 nm, as demonstrated in UV–Visible diffuse reflectance spectra (UV–Vis DRS), confirm the production of Ag NPs, and the band at 264 nm is due to the $n$–$\pi^*$ transition [51].

#### 9.4.1.3 Sensor implementation employing green Ag NPs for detecting ammonia

Ammonia is widely utilized in fertilizers, animal feed, fiber production, medicines, and other applications. Ammonium salts are also utilized as food additives and as cleaning agents. Water and alcohol rapidly dissolve ammonia and its salts, which causes severe health problems for humans and aquatic animals. As a result, ammonia detection, sensing, and monitoring in water are critical. Here, Ag NPs synthesized using the green route are used for sensing the presence of ammonia in water. The rate of synthesis of Ag NPs is influenced when different concentrations of ammonia are added to the mixture of leaf extract and silver nitrate ($AgNO_3$). The formation of Ag NPs increases with increasing ammonia concentration, as indicated by the change in the intensity of the yellow color, which proves that the presence of ammonia accelerates the formation of Ag NPs. The UV–Vis spectra of the Ag NPs obtained after 15 minutes, the intensity of the absorbance of Ag NPs is much lower in the absence of ammonia. The SPR absorbance intensity of Ag NPs is linearly correlated with the concentration of ammonia. The linear coefficient ($R^2$) decreases with increasing reaction time, confirming that the sensing of ammonia at a low reaction time (15 minutes) is better than that at a longer reaction time.

#### 9.4.1.4 Pesticide detection using green-synthesized Ag NPs

The sensing capacity of the neem leaf-based green-synthesized Ag NPs was also determined using mancozeb (MCZ) agro-fungicide. It was observed that with increasing MCZ concentrations, a blueshift in the SPR peak was detected, and the Ag NPs displayed a linear response with a rise in the concentration of MCZ. In addition, up to 1 mM MCZ, the intensity and FWHM (full width at half maximum) of the SPR peak decreased, but the intensity and FWHM of the SPR peak increased for $C_{MCZ}$ concentrations greater than 1 mM [49]. The particle size, shape, interparticle spacing, agglomeration, and surrounding dielectric media all contribute to the shift in the SPR peak [8, 9]. As the surrounding medium (MCZ) can influence the resultant electric field at the MNP surface, changes in the MCZ concentration can cause changes in the amplitude, FWHM, and position of the SPR peak of Ag NPs in this situation [9]. The decline in the plasmonic coupling of Ag NPs due to the interaction between the Ag NPs and MCZ molecules can be the reason for the decrease in amplitude and narrowing of the SPR peak up to $C_{MCZ}$ < 1 mM. Moreover, the electric field due to incident light will be minimally affected at $C_{MCZ}$ < 1 mM owing to the higher ratio of the number of nanoparticles to the number of MCZ molecules. As a result, the interparticle interaction between Ag NPs is reduced, and resonance conditions are altered, causing an increase in the total restoring force and resulting in a greater resonant frequency and a shift of the SPR band toward a lower wavelength region [9]. When $C_{MCZ}$ is greater than 1 mM, the ratio of NPs to MCZ molecules decreases, and the SPR signal from Ag NPs is filtered by the MCZ molecules. As the concentration of MCZ increases, so does its light absorption ability, and as a result, the absorbance rises. This demonstrates that at higher $C_{MCZ}$, the SPR band's signal becomes less effective. The ratio of the shift in the SPR peak ($\Delta\lambda$) with respect to the change in the MCZ concentration ($\Delta C_{MCZ}$) defines the sensitivity of the biosensor and is given by the mathematical equation sensitivity ($S$) = $|\Delta\lambda$ (nm)/ $\Delta C_{MCZ}$ (mM)$|$. The sensitivity determined as 39.1 nm/mM shows that the green-synthesized Ag NPs using neem leaf extract have high sensitivity and can thus be used in a variety of cost-effective and environmentally friendly sensing device applications.

## 9.5 Conclusions, Outlook and Perspectives

The development of nanotechnology with green chemistry methods has led to inevitability in developing environmentally friendly methods for the production of nontoxic and biocompatible nanoparticles that have unmatched significance in the field of nanotechnology. It was observed that the increasing popularity of green methods has triggered the synthesis of nanoparticles using various sources, such as microorganisms and plants, resulting in large-scale production with less contamination. Green synthesis of nanoparticles from plant extracts is gaining good response as the plant extracts are readily available. The simplicity of handling, availability and broad viability of metabolites are some of the advantages of using plant extracts. It was observed that the ZnO nanoparticles produced using Ihlamur leaf extract had a single-phase hexagonal structure with an average crystallite size of 22 nm. The sphere-shaped nature of the nanoparticles was revealed by SEM, with little aggregation. As the plant extract serves as a reducing and capping agent, green technology is more cost-effective and environmentally friendly than alternative ways to produce ZnO nanoparticles. Green $TiO_2$ nanoparticles with dimensions of 13 nm were synthesized using *B. orellana* seed extract. XRD examination revealed the formation of the pure anatase phase in green $TiO_2$ nanoparticles. The inclusion of a mesoporous structure in the green $TiO_2$ nanoparticles with enlarged pore sizes and increased surface area improves the solar cell power conversion efficiency. The performance of the $ZnO/TiO_2$ nanoparticle-based solar cell was evaluated using current density–voltage characteristics when sunlight falls on it. The solar cell using green $TiO_2$ nanoparticle films has an overall current conversion efficiency of 3% due to the high photocurrent density. The constructed DSSC offers high efficiency and is attributable to a considerable increase in dye molecule absorption onto the surface of the $ZnO/TiO_2$ nanoparticles. As a result, constructing DSSCs using green-synthesized nanoparticles is a simple and promising method for energy applications. All of the sensing techniques described here are ecologically safe and provide a cost-effective method of identifying dangerous compounds. Ag NPs produced with neem leaf extract solution showed a strong SPR band and displayed a high level of stability. The XRD spectrum revealed that the produced Ag NPs had a hexagonal crystal structure, and the SEM image proved that spherical Ag NPs had formed. The rate of creation of Ag NPs without ammonia is slower than that of other ammonia-containing solutions, as indicated by the color change in the solution from yellow to brown and validated by UV–Vis spectra. The concentration of dissolved ammonia was linearly proportional to the intensity of silver absorption, and the blueshift in the SPR peak with increased ammonia content was due to the size reduction. The sensing properties of Ag NPs were also investigated using MCZ fungicide, which showed a linear relationship with MCZ concentration. The green sensor systems presented in this chapter used a simple method for synthesizing effective nanoparticles for sensor implementation and are based mostly on molecules present on nanoparticle surfaces.

## Acknowledgments

Dr. Kaushik Pal is grateful to his associate colleagues at Chandigarh University, Punjab, from UCRD facilities. He owes his sincere thanks to his colleagues, including Bachelor/Masters students. He sincerely thanks Vice-Chancellor Prof. Sabu Thomas for utilizing IIUCN, School of Chemical Science Laboratory and School of Energy Materials, at Mahatma Gandhi University for providing excellent research foundation.

## Conflict of Interest

There is no conflict of interest to perform this work successfully.

## References

1. Su, C. and R.W. Puls, Kinetics of trichloroethene reduction by zerovalent iron and tin: pretreatment effect, apparent activation energy, and intermediate products, *Environmental Science and Technology*, 1999, **33**(1), pp. 163–168.
2. Ponder, S.M., J.G. Darab, and T.E. Mallouk, Remediation of Cr (VI) and Pb (II) aqueous solutions using supported, nanoscale zero-valent iron. *Environmental Science and Technology*, 2000, **34**(12), pp. 2564–2569.
3. Lee, C., et al., Bactericidal effect of zero-valent iron nanoparticles on *Escherichia coli*. *Environmental Science and Technology*, 2008. **42**(13): pp. 4927–4933.
4. García, A., et al., Acute toxicity of cerium oxide, titanium oxide and

iron oxide nanoparticles using standardized tests. *Desalination*, 2011. **269**(1–3): pp. 136–141.

5. Anastas, P.T., et al., The role of catalysis in the design, development, and implementation of green chemistry. *Catalysis Today*, 2000. **55**(1–2): pp. 11–22.
6. Lu, Y. and S. Ozcan, Green nanomaterials: On track for a sustainable future. *Nano Today*, 2015. **10**(4): pp. 417–420.
7. Mohanpuria, P., N.K. Rana, and S.K. Yadav, Biosynthesis of nanoparticles: technological concepts and future applications. *Journal of Nanoparticle Research*, 2008. **10**(3): pp. 507–517.
8. Mittal, A.K., Y. Chisti, and U.C. Banerjee, Synthesis of metallic nanoparticles using plant extracts. *Biotechnology Advances*, 2013. **31**(2): pp. 346–356.
9. Makarov, V., et al., "Green" nanotechnologies: synthesis of metal nanoparticles using plants, *Acta Naturae*, 2014, **6**(1), pp. 35–44.
10. Raveendran, P., J. Fu, and S.L. Wallen, Completely "green" synthesis and stabilization of metal nanoparticles. *Journal of the American Chemical Society*, 2003. **125**(46): pp. 13940–13941.
11. Shah, M., et al., Green synthesis of metallic nanoparticles via biological entities. *Materials*, 2015. **8**(11): pp. 7278–7308.
12. Ovais, M., et al., Biosynthesis of metal nanoparticles via microbial enzymes: a mechanistic approach. *International Journal of Molecular Sciences*, 2018. **19**(12): p. 4100.
13. Yusof, H.M., R. Mohamad, and U.H. Zaidan, Microbial synthesis of zinc oxide nanoparticles and their potential application as an antimicrobial agent and a feed supplement in animal industry: a review. *Journal of Animal Science and Biotechnology*, 2019. **10**(1): pp. 1–22.
14. Varma, R.S., Greener approach to nanomaterials and their sustainable applications. *Current Opinion in Chemical Engineering*, 2012. **1**(2): pp. 123–128.
15. Al-Shmgani, H.S., et al., Biosynthesis of silver nanoparticles from Catharanthus roseus leaf extract and assessing their antioxidant, antimicrobial, and wound-healing activities. *Artificial Cells, Nanomedicine, and Biotechnology*, 2017. **45**(6): pp. 1234–1240.
16. Anastas, P.T. and J.C. Warner, Green chemistry. *Frontiers*, 1998. **640**: p. 1998.
17. Naik, R.R., et al., Biomimetic synthesis and patterning of silver nanoparticles. *Nature Materials*, 2002. **1**(3): pp. 169–172.

18. Hakim, L.F., et al., Aggregation behavior of nanoparticles in fluidized beds. *Powder Technology*, 2005. **160**(3): pp. 149–160.

19. Ovais, M., et al., Current state and prospects of the phytosynthesized colloidal gold nanoparticles and their applications in cancer theranostics. *Applied Microbiology and Biotechnology*, 2017. **101**(9): pp. 3551–3565.

20. Ajitha, B., Y.A.K. Reddy, and P.S. Reddy, Green synthesis and characterization of silver nanoparticles using Lantana camara leaf extract. *Materials Science and Engineering*: C, 2015. **49**: pp. 373–381.

21. Dhillon, G.S., et al., Green approach for nanoparticle biosynthesis by fungi: current trends and applications. *Critical Reviews in Biotechnology*, 2012. **32**(1): pp. 49–73.

22. Mahendra, R., et al., Myconanotechnology: a new and emerging science. *Applied Mycology*, 2009: pp. 258–267.

23. Devatha, C.P. and A.K. Thalla, Green synthesis of nanomaterials, in *Synthesis of inorganic nanomaterials*, Bhagyaraj, S. M., Oluwafemi, O. S., Kalarikkal, N., Thomas, S., eds., 2018, pp. 169–184. Amsterdam: Elsevier.

24. Javaid, A., et al., Diversity of bacterial synthesis of silver nanoparticles. *BioNanoScience*, 2018. **8**(1): pp. 43–59.

25. Njagi, E.C., et al., Biosynthesis of iron and silver nanoparticles at room temperature using aqueous sorghum bran extracts. *Langmuir*, 2011. **27**(1): pp. 264–271.

26. Shahwan, T., et al., Green synthesis of iron nanoparticles and their application as a Fenton-like catalyst for the degradation of aqueous cationic and anionic dyes. *Chemical Engineering Journal*, 2011. **172**(1): pp. 258–266.

27. Kumar, K.M., et al., Biobased green method to synthesise palladium and iron nanoparticles using *Terminalia chebula* aqueous extract. Spectrochimica Acta Part A: *Molecular and Biomolecular Spectroscopy*, 2013. **102**: pp. 128–133.

28. Kuang, Y., et al., Heterogeneous Fenton-like oxidation of monochlorobenzene using green synthesis of iron nanoparticles. *Journal of Colloid and Interface Science*, 2013. **410**: pp. 67–73.

29. Thakur, S. and N. Karak, One-step approach to prepare magnetic iron oxide/reduced graphene oxide nanohybrid for efficient organic and inorganic pollutants removal. *Materials Chemistry and Physics*, 2014. **144**(3): pp. 425–432.

30. Senthil, M. and C. Ramesh, Biogenic synthesis of $Fe_3O_4$ nanoparticles using Tridax procumbens leaf extract and its antibacterial activity on Pseudomonas aeruginosa. *Digest Journal of Nanomaterials and Biostructures (DJNB)*, 2012. **7**(4): pp. 1655–1660.

31. Jia, L., et al., The biosynthesis of palladium nanoparticles by antioxidants in Gardenia jasminoides Ellis: long lifetime nanocatalysts for p-nitrotoluene hydrogenation. *Nanotechnology*, 2009. **20**(38): p. 385601.

32. Asiya, S., et al., Sustainable preparation of gold nanoparticles via green chemistry approach for biogenic applications. *Materials Today Chemistry*, 2020. **17**: p. 100327.

33. Ahsanulhaq, Q., A. Umar, and Y. Hahn, Growth of aligned ZnO nanorods and nanopencils on ZnO/Si in aqueous solution: growth mechanism and structural and optical properties. *Nanotechnology*, 2007. **18**(11): p. 115603.

34. Reddy, S., et al., Preparation of NiO/ZnO hybrid nanoparticles for electrochemical sensing of dopamine and uric acid. *Chemical Sensors*, 2012. **2**(1): 1082.

35. Shashanka, R., et al., Fabrication and characterization of green synthesized ZnO nanoparticle based dye-sensitized solar cells. *Journal of Science: Advanced Materials and Devices*, 2020. **5**(2): pp. 185–191.

36. Rajendrachari, S., et al., Antimicrobial investigation of CuO and ZnO nanoparticles prepared by a rapid combustion method. *Physical Chemistry Research*, 2019. **7**(4): pp. 799–812.

37. Rajendrachari, S. and K. BE, Biosynthesis of silver nanoparticles using leaves of Acacia melanoxylon and their application as dopamine and hydrogen peroxide sensors. *Physical Chemistry Research*, 2020. **8**(1): p. 1–18.

38. Cormier, P.-A., et al., Single crystalline-like and nanostructured $TiO_2$ photoanodes for dye sensitized solar cells synthesized by reactive magnetron sputtering at glancing angle. *Journal of Physical Chemistry C*, 2018. **122**(36): pp. 20661–20668.

39. Khade, G., et al., Green synthesis of $TiO_2$ and its photocatalytic activity. *Journal of Materials Science: Materials in Electronics*, 2015. **26**(5): pp. 3309–3315.

40. Maurya, I.C., et al., Green synthesis of $TiO_2$ nanoparticles using Bixa orellana seed extract and its application for solar cells. *Solar Energy*, 2019. **194**: pp. 952–958.

41. Kushwaha, R., et al., Synthesis and characterization of nitrogen-doped $TiO_2$ samples and their application as thin film electrodes in dye-sensitized solar cells. *Journal of Solid State Electrochemistry*, 2015. **19**(2): pp. 507–517.

42. Perumal, S., et al., Synthesis and characterization studies of solvothermally synthesized undoped and Ag-doped $TiO_2$ nanoparticles using toluene as a solvent. *Journal of Engineering Research and Applications*, 2014. **4**(7): pp. 184–187.

43. León, A., et al., FTIR and Raman characterization of $TiO_2$ nanoparticles coated with polyethylene glycol as carrier for 2-methoxyestradiol. *Applied Sciences*, 2017. **7**(1): p. 49.

44. Jain, S. and M.S. Mehata, Medicinal plant leaf extract and pure flavonoid mediated green synthesis of silver nanoparticles and their enhanced antibacterial property. *Scientific Reports*, 2017. **7**(1): pp. 1–13.

45. Kamakshi, K., et al., Tuning the surface plasmon resonance and surface-enhanced Raman scattering of pulsed laser deposited silver nanoparticle films by ambience and deposition temperature. *Journal of Optics*, 2014. **16**(5): p. 055002.

46. García, M.A., Surface plasmons in metallic nanoparticles: fundamentals and applications. *Journal of Physics D: Applied Physics*, 2011. **44**(28): p. 283001.

47. Verma, A. and M.S. Mehata, Controllable synthesis of silver nanoparticles using Neem leaves and their antimicrobial activity. *Journal of Radiation Research and Applied Sciences*, 2016. **9**(1): pp. 109–115.

48. Shinde, N., A. Lokhande, and C. Lokhande, A green synthesis method for large area silver thin film containing nanoparticles. *Journal of Photochemistry and Photobiology B: Biology*, 2014. **136**: pp. 19–25.

49. Varghese Alex, K., et al., Green synthesized Ag nanoparticles for bio-sensing and Photocatalytic applications. *ACS Omega*, 2020. **5**(22): p. 13123–13129.

50. Anandalakshmi, K., J. Venugobal, and V. Ramasamy, Characterization of silver nanoparticles by green synthesis method using Pedalium murex leaf extract and their antibacterial activity. *Applied Nanoscience*, 2016. **6**(3): pp. 399–408.

51. Xie, Y., et al., Plasmon-assisted site-selective growth of Ag nanotriangles and Ag-$Cu_2O$ hybrids. *Scientific Reports*, 2017. **7**(1): pp. 1–9.

## Chapter 10

# A New Hope to Green Nano-Biomedical Science and Technical Utilization

**Suma Sarojini, Sonam Dhargyal, Indhu Philip, Soma Biswas, Pruthviraj Ganesh Chavan, and Nachiketha Ramananda Rao**
*Department of Life Sciences, CHRIST (Deemed to be University), Bengaluru, Karnataka, India*
suma@christuniversity.in

The entire world is marching toward sustainable ways of life. In the medical field too, these options hold good. One such application is green nanotechnology (GNT), which uses nanoparticles (NPs) to target tumors, deliver drugs, betteren medical imaging, and so on. Some treatments based on NPs are also multifunctional; the same technique can be used to find tumors and carry drugs for treatment. This kind of targeted approach is of great significance, as the amount of drug injected into the body can be lower and does not pose much harm to healthy cells. If these NPs are made using GNT, they have the added advantage of not posing a threat to the environment in the process of their manufacture, use and disposal. These greener methods would definitely be a more sustainable approach than the chemical methods that are currently popular.

*Green Nanoarchitectonics*
Edited by Kaushik Pal

ISBN 978-981-4968-50-8 (Hardcover), 978-1-003-31860-6 (eBook)
www.jennystanford.com

Some promising approaches include dendrimers, liposomes, metal-organic frameworks, superparamagnetic NPs, and so on. Great knowledge has been achieved in recent years in unraveling the interactions of green nanomaterials with normal and cancer cells. This coupled with the understanding of nano-biointerfaces would catapult the research in this domain. A large number of plant extract NP combinations have shown positive results in cancer treatment. The current chapter highlights the latest technical developments of GNT in biomedical science and its optimal use in various disease diagnostic and therapeutic techniques without causing much toxicity. These greener methods would definitely be a more sustainable approach than the chemical methods that are currently popular and bring new hope in the field of nanomedicine.

## 10.1 Scope of Green Nanotechnology in Biomedical Science

In the past few decades, nanotechnology (NT) has found a niche in multiple domains of biomedical science, be it in diagnostics or therapeutics. Scientists are constantly searching for safer and more efficient nanomaterials to be employed in medicine. In recent years, green nanotechnology (GNT) has become an even better option, as it is environmentally sustainable and safer for patients. Reduced consumption of nonrenewable raw materials, better energy efficiency and lower greenhouse gas emissions make GNT a better option in the current scenario when the whole of planet Earth is highly polluted. Some of the fields wherein GNT has potential applications include the formulation of drugs with antibacterial, antioxidant, anti-inflammatory, antidiabetic, and anticancer properties. It also has huge scope in the field of drug delivery via dendrimers, liposomes, metal-organic frameworks (MOFs), superparamagnetic nanoparticles (NPs), and so on, and in biomedical imaging. Interactions of green nanomaterials with normal and cancer cells have been explored using extracts of many plants. Understanding nano-biointerfaces is also a prominent area in nanobiomedical science both in therapeutics and diagnostics. Various technical utilizations of GNT are discussed in the current chapter.

## 10.2 GNT in Diagnostics

*In vitro* diagnostics (IVD) include any tests performed using blood, serum, urine, tissue biopsy, and so on, to gain specific information about a patient's physiological condition. With the advancement of NT, the diagnostic domain has witnessed a tremendous transition, which can be observed in the use of nanoscale diagnostic kits as well as samples instead of bulk materials and large amounts of samples. Applications of NT are widely accepted because the large surface-to-volume ratio is so high, precise design of the devices, development of qualitatively distinct technologies, and special electrical and optical properties [1]. Studies on *in vitro* diagnostic approaches based on antibody immobilization have been attempted by many researchers. Target antigen detection through direct antibody immobilization in green synthesized chitosan-based gold NPs (Ab-CS-GNPs) is a novel *in vitro* diagnostic approach with the pros of low cost, increased functionality, and so on. Here, Ab-CS-GNPs are used in sandwich-like immunoassays where GNPs act as a label for the antibody [2].

### 10.2.1 Applications in Blood Glucose Monitoring

Glucose sensors are important *in vitro* diagnostic devices used to measure blood glucose levels as a part of managing diabetes mellitus. Different designs of eco-compatible NT-based glucose biosensors have been developed recently. One such discovery is the use of a green approach for making nanoporous gold NPs (npGNPs) using Au-Ag alloys. The reducing agent used for the synthesis of aqueous phase npGNPs was extracted from *Oryza sativa* (rice) seeds. The effectiveness of green synthesized npGNPs was exponentially higher than that of their metallic counterparts and has potential use in nonenzymatic glucose sensing [3]. Another nonenzymatic glucose biosensor was developed by using pencil graphite electrodes modified by copper NPs. It was for the first time where pencil graphite was converted to reduced graphene oxide (GO), and the reduction was done by the use of scanning potentials and thus became a green method. This biosensor is inexpensive, simple, and reproducible with good reliability in glucose sensing [4].

An amperometric glucose biosensor was developed from green synthesized zinc oxide (ZnO) NP using *Zingiber officinale* root.

Compared to previous ZnO NP biosensors, this is unique because of its detection limit, low cost, low $K_m$ value, response time, and so on. It possesses an anti-interference capacity against uric acid and ascorbic acid [5]. Silver-doped cerium oxide NPs synthesized from *Euphorbia hirta* ethanolic extract are a novel glucose biosensor that uses the cyclic voltammetry method to detect the glucose concentration [6]. Silver NPs (Ag NPs) produced using *Ocimum tenuiflorum* leaf extract were found to be effective nonenzymatic glucose biosensors. Ag NPs were applied onto glassy carbon electrodes (GCEs), which exhibited good anti-interference capacity. Such biosensors have fast action, high reproducibility and stability [7]. The process of green synthesized carbon nanostructures and carbon sheets is widely accepted in current times because of the environmental incompatibility and time-consuming process of conventional graphene synthesis. Based on the green synthesis method, unique, stable ZnO NPs are made and embedded onto nitrogen-doped carbon sheets. Peach extract, Zn powder, and aqueous ammonia were used as the precursors of carbon, ZnO, NP, and nitrogen, respectively. This enzymatic glucose sensor can be utilized for quantitative glucose detection in human blood serum [8]. A few examples of green nanoglucose biosensors are depicted in Fig. 10.1.

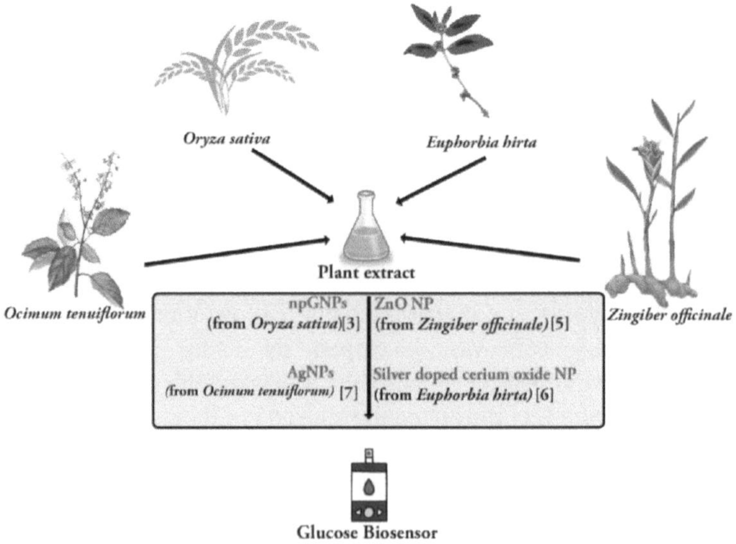


**Figure 10.1** Green nanobiosensors in blood glucose detection.

### 10.2.2 Applications in Cancer and Lifestyle Disease Diagnostics

There are very few reliable tumor markers used in oncological *in vitro* diagnostics. Recently, developed green synthesized magnetic iron oxide NPs for prostate-specific antigen detection (PSA) can be used as a potential substitute for ELISA kits. Magnetites used in this assay were green synthesized from blackberry leaf extract [9]. Metal organic framework PCN-224 and DNA tetrahedron-associated dual aptamer-based electrochemical cytosensors are capable of capturing MCF-7 cancer cells and are an efficient method for early breast cancer diagnostics [10]. Carbohydrate antigen (CA 19-9) is a pancreatic tumor marker, and its detection from biological samples can be used as a rapid diagnostic method for pancreatic cancer. Fluorescence immunosensing through carbon quantum dot (QD)-incorporated gold NPs was developed for the detection of CA 19-9. The synthesis was performed in an environmentally friendly green method via microwave assistance [11].

Diagnosis and monitoring of serum cholesterol levels are important because of the risk factors for heart diseases, prostate cancer, and so on. Palladium NPs were synthesized using gum kondagogu in a simple green method of autoclaving the reactants at 121°C for 30 min. The innate peroxidase activity of these NPs accompanied by oxidation of cholesterol by cholesterol oxidase leads to colorimetric oxidation, enabling cholesterol detection. This assay has potential application in clinical diagnosis because of its sensitivity [12]. Another advanced method for green synthesis of NP is through the use of sonochemical techniques. Bimetal oxide manganese ferrite ($MnFe_2O_4$) NPs synthesized using sonochemical methods are more effective in serum detection of the neurotransmitter 5-hydroxytryptamine [13].

### 10.2.3 Applications in Microbial Disease Diagnostics

Most of the detection techniques for microbial pathogens are time consuming and complex. Among bacterial diseases, listeriosis caused by the foodborne pathogen *Listeria monocytogenes* is catastrophic because of its high risk to pregnant women. Different techniques, such as immunoassays, culture-dependent methods and nucleic acid-based methods, are used for pathogen detection. Recently, a simple and rapid nanoscale liposome-based gold NP platform was developed for the efficient detection of listeriolysin O (LLO) toxin. LLO is a pore-forming toxin produced by the pathogen, and the detection is performed by observing a colorimetric reaction [14]. Another widely used detection technique for Listeria infection is lateral flow immunoassay (LFIA) based on gold NPs. Green synthesis of salt-tolerant gold NPs using Damask rose petal aqueous extract is a rapid and environmentally friendly *in vitro* diagnostic method. It has better applications, as the antibody–AEDR-Au NP LFIA strips are sensitive and distinct. Its sensitivity is pillared by detecting traceable concentrations of the pathogen, which is manifold less than the infection dose [15]. Examples of a few promising green nanobiosensors for microbial detection are depicted in Fig. 10.2.

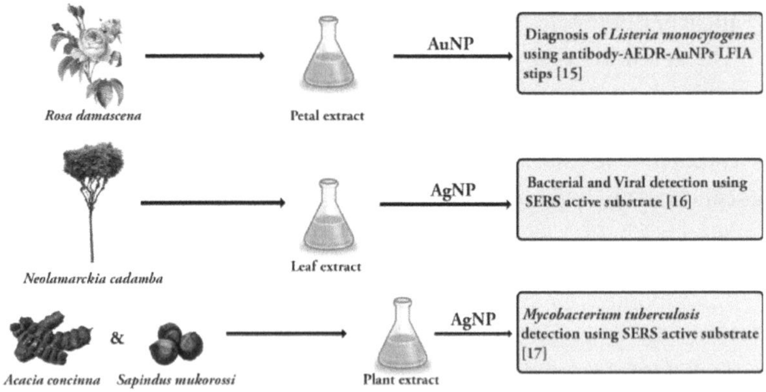


**Figure 10.2** Green nanobiosensors in microbial pathogen detection.

## 10.3 GNT in Biomedical Imaging

Biomedical imaging has become a foundation of modern medicine. It permits the visualization of cells and biological processes in living subjects noninvasively and the diagnosis of disease at early stages. Breakthroughs in imaging techniques have paved the way for the growth of novel, sensitive, and facultative contrasting agents. To date, numerous NPs have been developed as contrast agents due to their small size, controllable physical properties, high sensitivity, long circulation time, and facile surface modification. The surface of NPs is often conjugated with suitable targeting ligands. Multifunctional NPs have been developed, integrating various ligands. This enables therapy and multimodal imaging simultaneously, known as theranostics. X-ray computed tomography (CT), magnetic resonance imaging (MRI), and positron emission tomography (PET) are examples of molecular imaging techniques that allow observation of organs by the naked eye without disturbing them.

### 10.3.1 Applications in MRI

MRI is the most convenient imaging method available in clinical and research settings. Since MRI has low sensitivity, the relaxivity of the contrast agent should be very high. Liposomes and micelles, which are lipid-based NPs, have been extensively used as drug carrier vehicles. A potential application of lipidic NPs is their utility as a contrast agent in multimodal MRI. Lipids contain both hydrophobic and hydrophilic parts, which join to form aggregates in an aqueous environment. They arrange themselves in such a way that hydrophilic components face water, and hydrophobic parts cluster together. Various structures, such as liposomes, disks, and spherical micelles, are formed. Hydrophobic iron oxide particles, perfluorocarbons, and QDs can all be dissolved with this method. Likewise, MR-detectable and fluorescent amphiphilic molecules are fused inside lipidic NPs. In the same way, paramagnetic agents such as $MnCl_2$, Gd--DTPA (gadolinium--diethylenetriamine penta-acetic acid), and Mn-DTPA bound to serum proteins prepared in the liposomal formulation have been used as contrast agents for MRI. This type of contrast agent has been successfully used to detect tumors of the liver [18]. Another kind of liposome has been prepared where the paramagnetic molecules have been incorporated into the lipid bilayer. This technique resulted in better ionic relaxivity and is being used for the detection of tumors in lymph nodes [19]. Iron oxide NPs have been incorporated inside the liposomal lumen (magnetoliposomes), which has been used as a contrast agent in MRIs *in vivo* for the detection of bone marrow [20]. Micelles have been shown to be ideal candidates for use as MRI contrast agents; for example, gadofluorine, composed of a hydrophobic tail and a $Gd3^+$ hydrophilic head, has been successfully used for MR lymphography and atherosclerotic plaques [21]. High-density lipoprotein incorporated with Gd–DTPA has been used as a contrasting agent in detecting atherosclerosis [22]. Using single precursors of gold and iron and proanthocyanin from grape seed, gold and iron oxide multimodal NPs were prepared with magnetic and X-ray contrast properties. Such nanohybrids depicted superparamagnetism with high relaxivity and magnetic saturation. The simple combination of magnetite and gold showed better results in CT contrast as well as MRI compared to conventional iodine contrast agents. Even at larger doses and for longer periods of Cytotoxicity and oxidative stress were less even at extended times and increased dosages, thereby showing better biocompatibility [23].

### 10.3.2 Applications in X-Ray CT

CT is one of the most widely utilized noninvasive methods of diagnostics because of its deep tissue penetration, excellent 3D resolution, reconstruction capabilities and cost-effective nature. Green NP-based contrast agents were recently developed to overcome the drawbacks of iodine-based CT contrast agents, with better circulation half-lives and more X-ray absorption. More NPs with higher CT contrast must be delivered selectively to the target site to improve CT-based targeted imaging of specific tissues or cells, necessitating the development of new and effective targeted CT contrast agents with the ability to bind to specific receptors expressed by cells in the targeted tissues. Mannan is a bioactive polysaccharide with an $\alpha(1\rightarrow6)$-linked mannopyranose backbone that is extensively branched. Plants and some fungi have mannan in their cell walls. Mannan-based systems exhibit quicker identification and absorption via mannose receptors produced on the surface of antigen-presenting cells (APCs). Mannon-capped gold NPs produced by the green method and utilized in lymph node imaging are simple to make and nontoxic and have proven to be excellent contrast agents for CT imaging [24].

### 10.3.3 Applications in Optical Bioimaging

Optical bioimaging is based on optical contrast phenomena such as light reflection, transmission, and photoluminescence differences between the imaged region and the surrounding area. Semiconductor

QDs are currently the most attractive luminous nanomaterials due to their optoelectric and physicochemical characteristics. The use of nontoxic and environmentally friendly QD synthesis has increased in recent years. Among them, ZnS QDs stand out because of their biocompatibility, visible light emission, and capacity to accept a variety of capping agents. A single-pot method at different pH values has been successfully established for the green synthesis of ZnS QDs [25]. The optical and physicochemical properties of the nanocrystals are influenced by the surface capping agents. Nanomaterials have been created using biopolymers such as polysaccharides. Carboxymethylcellulose (CMC), a cellulose derivative, has been used as an intrinsically biocompatible, simple surface capping macromolecule for the production of NPs. The cell viability tests showed no cytotoxicity. As a result, tumor cells are being imaged at an early stage utilizing these ZnS fluorescent QDs stabilized by CMC at different pH values using a simple one-step green aqueous colloidal method [26]. Upconversion luminescence is a collection of nonlinear optical processes that convert two or more lower-energy pump photons (typically near IR light) into one higher-energy photon (UV, visible, and/or NIR light). Upconversion NPs show a long lifetime, high photostability, sharp emission bandwidth, and low cytotoxicity and hence are used for bioimaging applications, such as vascular imaging, tumor targeting, and cell tracking [27].

## 10.4 GNT in Regenerative Medicine

The ability of living tissue to autoregenerate eventually diminishes due to aging, thereby decreasing the quality and number of stem cells, the negative impact of inflammation on wound repair, or the low regenerative capacity of some tissues [28]. Regenerative medicine (RM) offers varied possibilities of repairing, regrowing and replacing cells, tissues and organs damaged due to trauma, cardiac diseases, aging, congenital defects, and so on [29]. The use of NT has been recognized as a potential therapeutic approach in RM. It involves a combination of scaffolds, growth factors and target cells to employ regeneration [30].

### 10.4.1 Polymer-Based GNT

The incorporation of modified halloysite nanotubes (mHNTs) into hydrogels can enhance the mechanical strength and lead to the multiplication and bone differentiation of mesenchymal stem cells encapsulated in the scaffold. This showed that the injectable chitosan hydrogel is a promising scaffold for bone regeneration [31]. The amalgamation of extracting green seaweed ulvan-cellulose with polydioxanone and polylactide electrospun into mats of nanofibers with a multitude of physicochemical characteristics scrutinized *in vitro* in a Wistar rat model using fibroblast cells did not show any foreign body response. Additionally, it showed increased angiogenesis for skin tissue regeneration [32]. Altering the surface of lignin nanofibers with arginine has been shown to encourage angiogenesis, collagen deposition, reepithelialization, and wound closure. This is due to the better wound healing process mediated by arginine and the structural similarity of the natural extracellular matrix with lignin nanofibers [33]. Biomineralization on the lignin nanofibrous surface by hydroxyapatite makes a perfect substrate for osteogenic conduction and cell proliferation. Relatively good osteoconductivity was observed in the case of composites of lignin with hydroxyapatite and polycaprolactone. This implied better potential for bone therapies [34]. Fabricated aligned electrospun polyvinyl alcohol–polyglycerol sebacate (PVA–PGS) fibers with varying percentages of 1%, 3%, and 5% wt lignin were prepared by employing the electrospinning technique. This has been shown to be efficient in the domain of nerve tissue engineering [35].

Nanofibrous chitosan/PCL can be used as a hepatoceles' scaffold to make the growth and attachment of hepatocytes. By increasing the hydrophilicity of the system by using the galacatosan method, the attachment ability of the HEP-G2 cell line could be improved. For liver regeneration, this scaffold was considered to be a promising substrate [36]. High conductivity, low toxicity, and excellent biocompatibility are shown by graphene-based materials. A phase separation approach was used to fabricate a piezoelectric GO-polyvinylidene fluoride (PVDF) network, whose electrical, mechanical and physical properties were correlated with the concentration of GO nanosheets. The water flux, water adsorption and hydrophilicity of the scaffolds were found to be increased with the introduction of GO nanosheets to the PVDF network. In addition, scaffolds treated with GO exhibited a piezoelectric β-phase fraction and electrical conductivity. The rate of proliferation of cells was found to be better than that of cells grown in non-GO-treated substrates [37].

### 10.4.2 Magnetic NPs in Tissue Engineering

Magnetic NPs can be used to alter cardiac scaffolds. Cardiosphere-derived stem cells were labeled using magnetic ferumoxytol NPs and were targeted into the coronary arteries of rat postmyocardial infarction ischemia/reperfusion models. This was found to increase the level of stem cell retention and engraftment, which normally pose a serious problem [38]. Magnetic ferrous oxide NPs coupled with antibodies were aimed at SIRPA (signal-regulatory protein alpha). This was found to be beneficial for the *in situ* manipulation of human cardiomyocytes with magnets [39]. In another study, a magnetic actuator was built using a silk fibroin matrix consisting of both the iron oxide NP and the drug. This complex could effectively stimulate the differentiation of embryonic H9C2 myoblast cells into cardiomyocytes [40]. To boost the multiplication of mouse embryonic cardiac cells, superparamagnetic iron oxide NPs (SPIONs) have been successfully added into silk fibroin nanofibers [41].

Magnetic scaffolds form the backbone of magnetic-tissue engineering (Mag-TE). A diluted combination of Matrigel and collagen was added to magnetically labeled cells using the Mag-TE platform, and this composite was cultured by incorporating polycarbonate pipes. Ultimately, this cell-polymer composite was identified as an ideal substrate for the regeneration of cardiac cells [42]. GNT can be applied to personalized medicine in the coming decade. It can revolutionize RM using stem cells from the same patient and deliver medicines to maintain a healthy environment for stem cells to grow and mature in the diseased tissues [43].

## 10.5 GNT in Drug Delivery

GNT has very efficient applications in drug delivery. It makes use of dendrimers, polymers, micelles, liposomes, carbon nanotubes, nanogels, porous silicon, and so on, which can be used for the preparation of NPs [44]. These NPs are more similar to carrier agents that take the drugs to the target organs. Sometimes drugs will not be able to reach the target organs due to the physiological effects of other organs and body parts; in cancer treatments, 99% of the medicines fail to reach their destined organs. Additionally, a high proportion is wasted. NP delivery will ensure that the drug is active until it reaches its destination. GNPs are economically suitable, with few or no side effects. In cancer treatment with GNPs, only cancerous cells will end up taking drugs and destroying themselves. This is another advantage over chemical therapies, which cause severe side effects. The short half-life in circulation, poor solubility, fast degradation and systemic toxicity could be overcome by GNT-based drug delivery systems. There are several studies of tailoring NPs to increase interactions with tissue microenvironments. This has been found to greatly improve (100- to 1000-fold) the site-specific delivery of NP-associated drugs compared to small molecules [45].

### 10.5.1 Silica-Based GNT

Many studies show that NPs made of porous silica are beneficial in controlled drug delivery. As the production of silica particles is laborious and involves harmful chemicals, this synthetic method of production can be replaced with more sustainable methods. Single-celled photosynthetic diatoms are the best option, and their microshells can be used for drug delivery because they have porosity at the nanoscale, biocompatibility, easy availability and low cost, which make them perfect drug delivery agents. Diatoms comprise a natural source of silica that can be produced in a relatively inexpensive way. Diatomaceous earth gets accumulated on the sea and lake beds by deposition of skeletons of dead diatoms over tens of lakhs of years. Their cell walls, called frustules, are laden with very pure silica [46].

### 10.5.2 Polysaccharide- and Lipid-Based GNT

Polysaccharides are also used for GNP-mediated drug delivery. Polysaccharides such as heparin are biocompatible, provide a large surface area for drug loading and are also cheap and efficient. GNPs produced using carbohydrates are used in the treatment of osteogenic disorders such as osteoporosis, bone tissue regeneration, and osteosarcomas [47]. The efficiency of cancer therapy can be greatly improved by targeting specific cancer sites using target ligands and lowering the side effects by the use of GNP anticancer drug combinations. Formulations of liposomal NPs (LNPs) with hydrophilic inner cores and hydrophobic lipid bilayers are made by incorporating the drug into one or more phospholipid bilayers in such a way that their size comes to only a few hundred nanometers. This kind of encapsulation greatly reduces the toxicity of drugs as the drug is distributed right on the target. LNPs accumulate in tumor cells by entering the cell membrane. One way of enhancing the half-life and targeting efficiency of LNPs is by surface modification using PEG. Coencapsulation of quercetin and vincistrine in a PEGylated liposome was attempted successfully for the treatment of breast carcinoma. This method helped in better and more prolonged drug circulation in plasma. It was also found to promote higher synergism and controlled release *in vivo* into JIMT-1 breast cancer cells. LNPs have been demonstrated to be efficient delivery models for peptides, oligonucleotides and siRNA-based gene therapy for various breast carcinomas owing to their lower cytotoxicity and higher silencing efficiency [48]. GNT has been extensively explored in killing cancer cells, a few of which are depicted in Fig. 10.3.

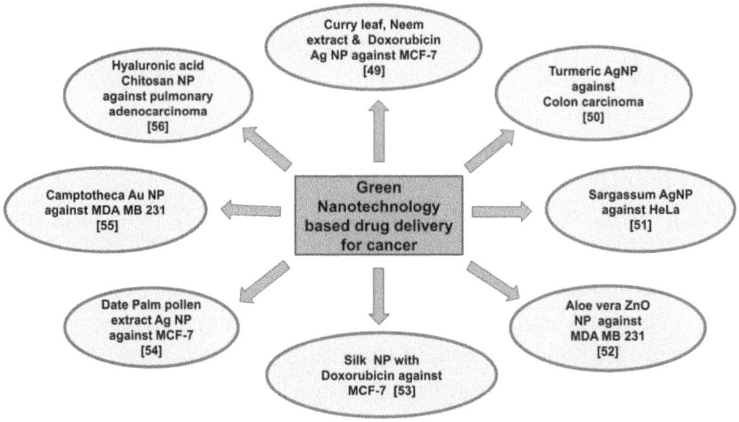


**Figure 10.3** Green nanotechnology formulations to treat different types of cancers.

## 10.6 Advantages, Limitations, and Future of GNT in Biomedicine

Green synthesis of NP reduces environmental toxicity and health-related complications. The process of green synthesis includes biological reagents and nonhazardous methods. Biological reagents are microorganisms and their enzymes or plant products such as isolates and extracts. Green synthesis methods confer various ascendancies over physical and chemical methods. These methods are generally very eco-friendly, do not involve the necessity for high pressure, energy and temperature, limit the usage of hazardous chemicals and are economically feasible. Green synthesis of Ag NPs from plant extracts is considered better due to their nonpathogenicity, nontoxic nature, easy availability and nonpolluting waste. The limitations of using biological synthesis are the high production cost for NPs and safety risks [57].

The biosynthesis of metal NPs (Mt NPs) using different microbes and microbial enzymes is cost-effective and environmentally friendly and has wide applications in biomedical NT. They also exhibit low toxicity, high stability and degradability. Due to their unique characteristics, microbial Mt NPs have found applications in the agriculture, medicine, cosmetics and biotechnology industries. Challenges that are involved in the biosynthesis of Mt NPs are selecting suitable biomolecules, maintaining the primary characteristics of the biosynthesized Mt NPs and optimizing different conditions suitable for their biosynthesis [58].

The biological and photocatalytic activities of metal oxide NPs are outlined as far better than those of metal NPs. Among all other metal oxide NPs, copper oxide NPs (CuO NPs) with anticancer, antimicrobial and antioxidant activity have proven to be a potent tool in biomedical applications. The synthesis of CuO NPs by physical and chemical methods requires labor intensiveness, complex procedures, high energy and the use of expensive and hazardous chemicals. Hence, the development of a biocompatible approach for the synthesis of CuO NPs is important. Plant-based synthesis of CuO NPs is gaining much attention because of its abundance, low energy requirement, the presence of different biomolecules, enhanced stability, cost-effectiveness, absence of hazardous chemicals and easy methods or procedures. The major limitation of the biological synthesis of metal oxide NPs using microorganisms is the safety risk, the need for aseptic cultivation and the high cost of production [59]. Studies have revealed that ZnO has been thoroughly studied as a prospective contender in biomedical applications. The use of GNT in the synthesis of ZnO NPs has also proven to be cost-effective and environmentally friendly. The biomedical use of green-synthesized ZnO NPs includes antimicrobial features, anticancer properties, tissue engineering, wound healing and biosensing capabilities. The lack of standardized biocompatible protocols and the possibility of contamination of the final NP in the case of microorganism-mediated synthesis are some of the limitations of their biomedical use. Hence, more research needs to be done to overcome the drawbacks of green synthesis [60]. Green-synthesized silver (Ag) NPs, ZnO NPs and mixed Ag/ZnO NPs were used for the production of clinical antimicrobial wound-healing bandages. Bandages impregnated with these NPs showed increased inhibition of bacterial growth. These biostable NP treatments are quite inexpensive, less toxic and effective against a wide range of microbes, which makes them an efficient tool in biomedical science. To understand the viable detrimental effects of these NPs on both humans and the environment, more clinical trials must be carried out [61]. Based on ongoing studies, SPIONs have been found to play a pivotal role in drug delivery, cancer therapy, Alzheimer's disease therapy and photodynamic therapy. SPIONs are made of tiny crystals of iron oxide and exhibit magnetic properties. The main advantage of using SPIONs in biomedicine is their biostability, multifunctionality, simple process of synthesis and scope for modifications. There are some negative effects of using SPIONs, including inflammatory responses, cytotoxicity, oxidative stress and epigenetic alterations [62].

GNT-guided drug delivery systems have helped to develop drug delivery vehicles such as metallic NPs (NPs), QDs, dendrimers, organic polymeric NPs, solid lipid NPs, nanomicelles, liposomes, and nanohydrogels, thus making them an efficient tool in nanomedicine. Metallic NPs such as Ag, Au, Pt, Cu, Fe, ZnO, and so on, have significant advantages because of their large surface area, which allows the binding or entrapment of drugs to the surface of the target molecule. Drug delivery-based NPs are potential drug delivery carriers or drug delivery matrices. Apart from the use of

NPs in drug delivery, they have also gained attention in biomedical imaging, such as fluorophores, QDs, gold NPs and iron oxide NPs. All these NPs are biodegradable and biocompatible polymers, which makes them efficient for use in pharmaceuticals, cosmetics, and so on. The potential use of drug delivery-based NPs as nanocarriers and nanomedicines has yet to be fully explored [63]. Using GNT, researchers are now trying to make "Nano-Ayurvedic medicines" such as NSB (gold NP-based Nano Swarna Bhasma) drugs by combining gold NPs and phytochemicals from mango peels. The antitumor properties of the NSB drug were checked in breast cancer cells. NSB was found to be cytotoxic to tumor cells and nontoxic to healthy cells. Hence, nano-Ayurvedic gold NP-based drugs are safe, exhibit great efficacy and can be used as adjuvant therapy to reduce the adverse effects of chemotherapy [64].

## 10.7 Conclusions, Outlook, and Future Prospects

Green nanotechnology is in its juvenile stage. Only a few nanomedicines exist that successfully utilize GNT. One of the lacunae might be the lack of well-defined parameters for pairing the correct ratio or combination of NP with the drug of interest. Although GNT offers a nontoxic and eco-friendly technique for the production of metallic NPs, the protocols for optimal production and scale up need to be modified further. Some of the parameters include optimizing the size, crystal growth, morphology, stability and aggregation of GNPs and their separation and purification. This has to be standardized for cost-effective production to be on par with traditional methods. GNP produced using plant extracts are generally stable. In the future, genetic engineering approaches can also be employed to produce a larger quantity of proteins and other biomolecules for better production of GNPs. Every technology has its own sets of advantages and limitations. The focus should be on eliminating the limitations of ongoing research, thus making GNT a very efficient tool in biomedical research. Therefore, researchers should focus more on optimizing the large-scale production of quality drugs and efficient delivery methods using GNT. Once it is nurtured in the best possible way, overcoming all the limitations and making it more adaptable, it will play a pivotal role in biomedicine.

In the coming years, this would be more of a rule than a choice, as we need to embrace sustainability in all walks of life.

## Acknowledgments

The authors are grateful to the faculty of the Department of Life Sciences, CHRIST (Deemed to be University), Bengaluru, for the support extended for the successful completion of this work.

# References

1. Li, Z., and Shum, H. C. (2019). Nanotechnology and microfluidics for biosensing and biophysical property assessment: Implications for next-generation *in vitro* diagnostics. *Nanotechnology for Microfluidics, 3,* 83–107.
2. Majdi, H., Salehi, R., Pourhassan-Moghaddam, M., Mahmoodi, S., Poursalehi, Z., and Vasilescu, S. (2019). Antibody conjugated green synthesized chitosan-gold nanoparticles for optical biosensing. *Colloids and Interface Science Communications, 33,* 100207.
3. Verma, N. (2018). A green synthetic approach for size tunable nanoporous gold nanoparticles and its glucose sensing application. *Applied Surface Science, 462,* 753–759.
4. Pourbeyram, S., Abdollahpour, J., and Soltanpour, M. (2019). Green synthesis of copper oxide nanoparticles decorated reduced graphene oxide for high sensitive detection of glucose. *Materials Science and Engineering C, 94,* 850–857.
5. Dönmez, S. (2020). Green synthesis of zinc oxide nanoparticles using Zingiber officinale root extract and their applications in glucose biosensor. *El-Cezeri Journal of Science and Engineering, 7*(3), 1191–1200.
6. Vennila, R., Hasina Banu, A., Kamaraj, P., Devikala, S., Arthanareeswari, M., Selvi, J. A., Pushpamalini, T., Buela, J. G., Priya, D., and Sivasankari, R. (2018). A novel glucose sensor using green synthesized Ag doped $CeO_2$ nanoparticles. *Materials Today: Proceedings, 5*(2), 8683–8690.
7. Dayakar, T., Venkateswara Rao, K., Park, J., Sadasivuni, K. K., Ramachandra Rao, K., and Jaya rambabu, N. (2018). Non-enzymatic biosensing of glucose based on silver nanoparticles synthesized from Ocimum tenuiflorum leaf extract and silver nitrate. *Materials Chemistry and Physics, 216,* 502–507.

8. Muthuchamy, N., Atchudan, R., Edison, T. N. J. I., Perumal, S., and Lee, Y. R. (2018). High-performance glucose biosensor based on green synthesized zinc oxide nanoparticle embedded nitrogen-doped carbon sheet. *Journal of Electroanalytical Chemistry, 816*, 195–204.
9. Muhd, B, K., Umar A., Muhammad, Y., Kani,Y.A., Iliya, S., Wali, U., Tahiru, A.,Abubakar, U.F., Shehu, Z., Ahmed, A.Y and Zainab, I. (2020). Ex vivo determination of prostate specific antigen. *International Journal of Research and Scientific Innovation, 7*, 296–302.
10. Ou, D., Sun, D., Liang, Z., Chen, B., Lin, X., and Chen, Z. (2019). A novel cytosensor for capture, detection and release of breast cancer cells based on metal organic framework PCN-224 and DNA tetrahedron linked dual-aptamer. *Sensors and Actuators, B: Chemical, 285*, 398–404.
11. Alarfaj, N. A., El-Tohamy, M. F., and Oraby, H. F. (2018). CA 19-9 pancreatic tumor marker fluorescence immunosensing detection via immobilized carbon quantum dots conjugated gold nanocomposite. *International Journal of Molecular Sciences, 19*(4), 1162.
12. Rastogi, L., Dash, K., and Sashidhar, R. B. (2021). Selective and sensitive detection of cholesterol using intrinsic peroxidase-like activity of biogenic palladium nanoparticles. *Current Research in Biotechnology, 3*, 42–48.
13. Elshikh, M. S., Chen, T. W., Mani, G., Chen, S. M., Huang, P. J., Ali, M. A., Al-Hemaid, F. M., and Al-Mohaimeed, A. M. (2021). Green sonochemical synthesis and fabrication of cubic MnFe2O4 electrocatalyst decorated carbon nitride nanohybrid for neurotransmitter detection in serum samples. *Ultrasonics Sonochemistry, 70*, 105305.
14. Mazur, F., Tran, H., Kuchel, R. P., and Chandrawati, R. (2020). Rapid detection of listeriolysin O toxin based on a nanoscale liposome-gold nanoparticle platform. *ACS Applied Nano Materials, 3*(7), 7270–7280.
15. Du, J., Wu, S., Hu, Z., Yu, Z., Zhao, D., and Bai, Y. (2020). Green synthesis of salt-tolerant gold nanoparticles for the rapid qualitative detection of *Listeria monocytogenes* in lateral flow immunoassay. *Journal of Materials Science, 55*(32), 15426–15438.
16. Ankamwar, B., Sur, U. K., and Das, P. (2016). SERS study of bacteria using biosynthesized silver nanoparticles as the SERS substrate. *Analytical Methods, 8*(11), 2335–2340. https://doi.org/10.1039/c5ay03014e
17. Kumar Sur, U., Ankamwar, B., Karmakar, S., Halder, A., and Das, P. (2018). Green synthesis of silver nanoparticles using the plant extract of shikakai and reetha. *Materials Today: Proceedings, 5*(1), 2321–2329. https://doi.org/10.1016/j.matpr.2017.09.236

18. Niesman, M. R., Bacic, G. G., Wright, S. M., Swartz, H. J., and Magin, R. L. (1990). Liposome encapsulated $MnCl_2$ as a liver specific contrast agent for magnetic resonance imaging. *Investigative Radiology*, *25*(5), 545–551.

19. Kabalka, G. W., Davis, M. A., Buonocore, E., Hubner, K., Holmberg, E., and Huang, L. (1990). Gd-labeled liposomes containing amphipathic agents for magnetic resonance imaging. *Investigative Radiology, 25*, S63–S64

20. Bulte, J.W., and Cuyper, M.D. (2003) Magnetoliposomes as contrast agents. *Methods in* Enzymology, *373*, 175–198.

21. Shimada, M., Yoshikawa, K., Suganuma, T., Kayanuma, H., Inoue, Y., Ito, K., and Hayashi, S. (2003). Interstitial magnetic resonance lymphography: Comparative animal study of gadofluorine 8 and gadolinium diethylenetriamine-pentaacetic acid. *Journal of Computer Assisted Tomography*, *27*(4), 641–646.

22. Frias, J. C., Williams, K. J., Fisher, E. A., and Fayad, Z. A. (2004). Recombinant HDL-like NP: A specific contrast agent for MRI of atherosclerotic plaques. *Journal of the American Chemical Society,* 126(50), 16316–16317.

23. Narayanan, S., Sathy, B. N., Mony, U., Koyakutty, M., Nair, S. V., and Menon, D. (2011). Biocompatible magnetite/gold nanohybrid contrast agents via green chemistry for MRI and CT bioimaging. *ACS Applied Materials and Interfaces*, 4(1), 251–260.

24. Uthaman, S., Kim, H. S., Revuri, V., Min, J.-J., Lee, Y., Huh, K. M., and Park, I.-K. (2018). Green synthesis of bioactive polysaccharide-capped gold NP for lymph node CT imaging. *Carbohydrate Polymers, 181*, 27–33.

25. Mansur, A. A., Mansur, H. S., Mansur, R. L., de Carvalho, F. G., and Carvalho, S. M. (2018) Bioengineered II–VI semiconductor quantum dot–carboxymethylcellulose nanoconjugates as multifunctional fluorescent nanoprobes for bioimaging live cells, *Spectrochimica Acta Part A*, *189*, 393–404.

26. Caires, A.J., Mansur, A. P., Carvalho, I. C., Carvalho, S. M., and Mansur, H. S. (2020). Green synthesis of ZnS quantum dot/biopolymer photoluminescent nanoprobes for bioimaging brain cancer cells. *Materials Chemistry and Physics*, *244*,122716.

27. Pu, Y., Leng, J., Wang, D., Wang, J., Foster, N. R., and Chen, J. (2018). Recent progress in the green synthesis of rare-earth doped upconversion nanophosphors for optical bioimaging from cells to animals. *Chinese Journal of Chemical Engineering*, *26*(10), 2206–2218.

28. Place, E. S., Evans, N. D., and Stevens, M. M. (2009). Complexity in biomaterials for tissue engineering. *Nature Materials*, *8*(6), 457–470.
29. Sahle, F. F., Kim, S., Niloy, K. K., Tahia, F., Fili, C. V., Cooper, E., Hamilton, D. J., and Lowe, T. L. (2019). Nanotechnology in regenerative ophthalmology. *Advanced Drug Delivery Reviews*, *148*, 290–307.
30. Harrison, R. H., St-Pierre, J. P., and Stevens, M. M. (2014). Tissue engineering and regenerative medicine: A year in review. *Tissue Engineering Part B: Reviews*, *20*(1), 1–16.
31. Kazemi-Aghdam, F., Jahed, V., Dehghan-Niri, M., Ganji, F., and Vasheghani-Farahani, E. (2021). Injectable chitosan hydrogel embedding modified halloysite nanotubes for bone tissue engineering. *Carbohydrate Polymers 269*, 1–9.
32. Madub, K., Goonoo, N., Gimié, F., Arsa, I.A., Schönherr, H., and Bhaw-Luximon, A. (2021). Green seaweed ulvan-cellulose scaffolds enhance in vitro cell growth and in vivo angiogenesis for skin tissue engineering. *Carbohydrate Polymers*, *251*, 1–12.
33. Reesi, F., Minaiyan, M., and Taheri, A. (2018). A novel lignin-based nanofibrous dressing containing arginine for wound-healing applications. *Drug Delivery and Translational Research*, *8*(1), 111–122.
34. Kim, M., Yeo, M., Kim, M., and Kim, G. (2018). Biomimetic cellulose/calcium-deficient-hydroxyapatite composite scaffolds fabricated using an electric field for bone tissue engineering. *RSC Advances*, *8*(37), 20637–20647.
35. Saudi, A., Amini, S., Amirpour, N., Kazemi, M., Kharazi, A. Z., Salehi, H., and Rafienia, M. (2019). Promoting neural cell proliferation and differentiation by incorporating lignin into electrospun poly (vinyl alcohol) and poly (glycerol sebacate) fibers. *Materials Science and Engineering: C*, *104*, 110005.
36. Ghahremanzadeh, F., Alihosseini, F., and Semnani, D. (2021). Investigation and comparison of new galactosylation methods on PCL/chitosan scaffolds for enhanced liver tissue engineering. *International Journal of Biological Macromolecules*, *174*, 278–288.
37. Abzan, N., Kharaziha, M., and Labbaf, S. (2019). Development of three-dimensional piezoelectric polyvinylidene fluoride-graphene oxide scaffold by non-solvent induced phase separation method for nerve tissue engineering. *Materials and Design*, *167*, 1–12.
38. Vandergriff, A. C., Hensley, T. M., Henry, E. T., Shen, D. L., Anthony, S., Zhang J. Y., and Cheng, K. (2014). Magnetic targeting of cardiosphere-derived stem cells with ferumoxytol nanoparticles for treating rats with myocardial infarction. *Biomaterials*, *35*, 8528–8539.

39. Zwi-Dantsis, L., Wang, B., Marijon, C., Zonetti, S., Ferrini, A., Massi, L., Stuckey, D. J., Terracciano, C. M., and Stevens, M. M. (2020). Remote magnetic nanoparticle manipulation enables the dynamic patterning of cardiac tissues. *Advanced Materials, 32*, 1904598.

40. Chouhan, D., Mehrotra, S., Majumder, O., and Mandal, B. B. (2019). Magnetic actuator device assisted modulation of cellular behavior and tuning of drug release on silk platform. *ACS Biomaterials Science and Engineering, 5*, 92–105.

41. Nazari, H., Heirani-Tabasi, A., Hajiabbas, M., Bani, M. S., Nazari, M., Mahabadi, V. P., Rad, I., Kehtari M., Tafti, S. H. A., and Soleimani, M. (2020). Incorporation of spion-casein core-shells into silk-fibroin nanofibers for cardiac tissue engineering. *Journal of Cellular Biochemistry, 121*, 2981–2993.

42. Matos, A. M., Gonçalves, A. I., El Haj, A. J., and Gomes, M. E. (2020). Magnetic biomaterials and nanoinstructive tools as mediators of tendon mechanotransduction. *Nanoscale Advances, 2*(1), 140–148.

43. Anitua, E., Prado, R., Troya, M., Zalduendo, M., de la Fuente, M., Pino, A., Muruzabal, F., and Orive, G. (2016). Implementation of a more physiological plasma rich in growth factor (PRGF) protocol: Anticoagulant removal and reduction in activator concentration. *Platelets, 27*(5), 459–466.

44. Slowing, I. I., Vivero-Escoto, J. J., Wu, C. W., and Lin, V. S. (2008). Mesoporous silica nanoparticles as controlled release drug delivery and gene transfection carriers. *Advanced Drug Delivery Reviews, 60*, 1278–1288.

45. Wolfram, J., Nizzero, S., Liu, H., Li, F., Zhang, G., Li, Z., Shen H., Blanco E., and Ferrari E. (2017). A chloroquine-induced macrophage-preconditioning strategy for improved nanodelivery. *Scientific Reports, 7*, 13738.

46. De Stefano, L., De Stefano, M., De Tommasi, E., Rea, I., and Rendina, I. (2016). A natural source of porous biosilica for nanotech applications: The diatoms microalgae. *Physica Status Solidi, 8*(6), 1820–1825.

47. Medina-Cruz, D., Mostafavi, E., Vernet-Crua, A., Cheng, J., Shah, V., Cholula-Diaz, J. L., Guisbiers, G., Tao, J., García-Martín, J. M., and Webster, T. J. (2020). Green nanotechnology-based drug delivery systems for osteogenic disorders. *Expert Opinion on Drug Delivery, 17*(3), 341–356.

48. Liyanage, P. Y., Hettiarachchi, S. D., Zhou, Y., Ouhtit, A., Seven, E. S., Oztan, C. Y., Celik, E., and Leblanc, R. M. (2019). Nanoparticle-mediated targeted drug delivery for breast cancer treatment, *Biochimica et Biophysica Acta (BBA)—Reviews on Cancer,* 1871(2), 419–433.

49. Mehnath, S., Arjama, M., Rajan, M., Arokia, M., and Jeyaraj, M. (2018). Polyorganophosphazene stabilized gold nanoparticles for intracellular drug delivery in breast carcinoma cells. *Process Biochemistry, 72,* 152–161.
50. Venkatadri, B., Shanparvish, E., Rameshkumar, M. R., Arasu, M. V., Al-Dhabi, N. A., Ponnusamy, V. K., and Agastian, P. (2020). Green synthesis of silver nanoparticles using aqueous rhizome extract of *Zingiber officinale* and *Curcuma longa*: In vitro anti-cancer potential on human colon carcinoma HT-29 cells. *Saudi Journal of Biological Sciences, 27*(11), 2980–2986.
51. Balaraman, P., Balasubramanian, B., Kaliannan, D., Durai, M., Kamyab, H., Park, S., Chelliapan, S., Lee, C. T., Maluventhen, V., and Maruthupandian, A. (2020). Phyco-synthesis of silver nanoparticles mediated from marine algae *Sargassum myriocystum* and its potential biological and environmental applications. *Waste Biomass Valorization, 11*(10), 11.
52. Batool, M., Khurshid, S., Daoush, W. M., Siddique, S. A., and Nadeem T. (2021). Green synthesis and biomedical applications of ZnO nanoparticles: Role of PEGylated-ZnO nanoparticles as doxorubicin drug carrier against MDA-MB-231(TNBC) cells line. *Crystals, 11*(4), 344.
53. Wongpinyochit, T., Uhlmann, P., Urquhart, A. J., and Seib F. P. (2015). PEGylated silk nanoparticles for anticancer drug delivery. *Biomacromolecules 16*(11), 3712–3722.
54. Banu, H., Renuka, N., Faheem, S. M., Ismail, R., Singh, V., Saadatmand, Z., Khan, S. S., Narayanan, K., Raheem, A., and Premkumar, K. (2018). Gold and silver nanoparticles biomimetically synthesized using date palm pollen extract-induce apoptosis and regulate P53 and Bcl-2 expression in human breast adenocarcinoma cells. *Biological Trace Element Research*, 186, 122–134.
55. Thirumurugan, A., Blessy, V., and Karthikeyan, M. (2018). Comparative study on doxorubicin loaded metallic nanoparticles in drug delivery against MCF-7-cell line, *Applications of Nanomaterial, 10,* 303–313.
56. Wang, T., Hou, J.H., Su, C., Zhao, L., and Shi, Y.(2017). Hyaluronic acid-coated chitosan nanoparticles induce ROS-mediated tumor cell apoptosis and enhance antitumor efficiency by targeted drug delivery via CD44. *Journal of Nanobiotechnology, 15*(7), 7.
57. Kaabipour, S., and Hemmati, S. (2021). A review on the green and sustainable synthesis of silver nanoparticles and one-dimensional silver nanostructures. *Beilstein Journal of Nanotechnology, 12,* 102–136.

58. Ovais, M., Khalil, A. T., Ayaz, M., Ahmad, I., Nethi, S. K., and Mukherjee, S. (2018). Biosynthesis of metal nanoparticles via microbial enzymes : A mechanistic approach. *International Journal of Molecular Sciences, 19*, 4100.

59. Akintelu, S. A., Similoluwa, A., Folorunso, F. A., and Oyebamiji, A. (2020). Green synthesis of copper oxide nanoparticles for biomedical application and environmental remediation. *Heliyon*, 6, e04508.

60. Cruz, D. M., Mostafavi, E., Vernet-crua, A., Barabadi, H., Shah, V., and Jorge, L. (2020). Green nanotechnology-based zinc oxide (ZnO) nanomaterials for biomedical applications : A review green nanotechnology-based zinc oxide (ZnO) nanomaterials for biomedical applications . *Journal of Physics: Materials, 3*, 034005.

61. Khatami, M., Varma, R. S., Zafarnia, N., Yaghoobi, H., Sarani, M., and Kumar, V. G. (2018). Applications of green synthesized Ag, ZnO and Ag/ZnO nanoparticles for making clinical antimicrobial wound-healing bandages. *Sustainable Chemistry and Pharmacy, 10*, 9–15.

62. Dulinska-Litewka, J., Łazarczyk, A., Hałubiec, P., Szafranski, O., Karnas, K., and Karewicz, A. (2019). Superparamagnetic iron oxide nanoparticles: Current and prospective medical applications. *Materials*, 12, 617.

63. Kanwar, R., Rathee, J., Salunke, D. B., and Mehta, S. K., (2019). Green nanotechnology-driven drug delivery assemblies. *ACS Omega, 4*, 8804–8815.

64. Khoobchandani, M., Katti, K. K., Thipe, V. C., and Srisrimal D. (2020). New approaches in breast cancer therapy through green nanotechnology and nano-ayurvedic medicine: Pre-clinical and pilot human clinical investigations. *International Journal of Nanomedicine*, 15, 181–197.

# Chapter 11

# Organometallic Nanomaterials Synthesis and Sustainable Green Nanotechnology Applications

**Mohammad Harun-Ur-Rashid[a] and Abu Bin Imran[b]**
[a]*Department of Chemistry,*
*International University of Business Agriculture and Technology, Dhaka, Bangladesh*
[b]*Department of Chemistry, Bangladesh University of Engineering and Technology, Dhaka, Bangladesh*
abimran@chem.buet.ac.bd

Organometallic nanomaterials, comprising metal epicenter organic linkers, are a novel category of materials with minute interstices. These outstanding and revolutionary hybrid nanomaterials have strived to have a significant impact on various research areas and industrial sectors, such as catalysis, energy transformation and storage, drug delivery, and sensing, because of their ultrahigh porosity, enormous internal surface area, crystallinity, flexibility of constituent geometry, structural diversity, adaptability, and various functionalities. Although the developments in this discipline are impressive, many challenges are required to be settled before the commercial ventures and applications of these materials, such as health risks, environmental impact, fabrication costs, and

*Green Nanoarchitectonics*
Edited by Kaushik Pal

ISBN 978-981-4968-50-8 (Hardcover), 978-1-003-31860-6 (eBook)
www.jennystanford.com

sustainability. Implementing sustainable green nanotechnology applications of organometallic nanomaterials would be the best way to resolve the present confrontations using environmentally benign solvents, reduce energy consumption, select renewable feedstock, eliminate wastage sources and design high-performance materials through theoretical predictions. The particular focus of this chapter is to carry out a systematic study in detail of the previous and present trends, status, and perspectives of synthetic routes for organometallic hybrid nanomaterials, including their background, composition, fabrication processes, and promising applications in sustainable green nanotechnology products. An inclusive analysis with a specific concentration on the advantages, drawbacks, and prospects of organometallic hybrid nanomaterials synthesized, especially for green and sustainable nanotechnology applications, will be addressed in this chapter.

## 11.1 Introduction

Organometallic nanomaterials are porous coordination systems belonging to the most advanced materials and are composed of metallic centers and organic linkers. The most commonly preferable organic linkers are aromatic heterocyclic or polyaromatic compounds, polytopic carboxylates, and imidazolate or pyridyl moieties that can ease coordination binding with selected metal ions. Metal ions from respective metal salts are usually used for synthesis, and counterions associated with the metal ions may be selected to address safety issues by lowering toxicity. Organometallic nanomaterial-based substructures are considered multifunctional crossover materials with outstanding characteristic advantages of inorganic metal particles, natural linkers, controllable porosity, and divergent usefulness. They offer structural tailorability, better porosity, large surface area, crystallinity, functional variation, and adaptability. Selective incorporation of intrinsic organometallic nanomaterials with well-defined functionalities can impart attractive features by collective consequences that excel the components. Most organometallic nanomaterials are commonly manufactured through solvothermal methods that negatively impact health and environmental components due to the utilization of toxic solvents, the involvement of long process times, and the consumption of a great deal of utilities. Although developments in this field are noteworthy, several limitations, challenges, and constraints, such as toxicity, efficacy, safety, instability, nonbiodegradable nature, manufacturing costs, high energy consumption, eco-toxicity, and environmental concerns, exist [1]. Several advancements in organometallic substructures have been studied and analyzed for their modern manufacturing methods [2], functional artificial light-harvesting systems [3], and catalytic performances [4, 5]. Apart from such attempts, our chapter concentrates on green synthetic routes for organometallic nanomaterials and their applications in sustainable applications in green nanotechnology with an additional focus on future prospects.

## 11.2 Synthesis of Organometallic Nanomaterials

Sustainable approaches to the synthesis and application of organometallic nanomaterials cover various strategies, including the selection of appropriate ligands and metal ions, utilization of benign or lowest amounts of solvents, and choosing the most effective techniques for green and sustainable synthesis [6]. Synthetic methods depend on the selection of ligands and metal ions. Green ligands for sustainable green synthesis of organometallic materials usually belong to three major categories: different types of aliphatic diacids, amino acids, and cyclodextrins (α-, β-, and γ-cyclodextrin). Metallic ions of s-block metals such as $Ca^{2+}$, $Mg^{2+}$, and $Fe^{3+}$, d-block metals such as $Zn^{2+}$, $Ti^{4+}$, and $Zr^{4+}$, and p-block metals including $Al^{3+}$ could be suitable candidates for green synthesis. The selection of solvents for the green synthesis of organometallic nanomaterials should be accomplished with due consideration of several parameters, such as process, worker, and environmental safety [7–9]. Water, isopropyl alcohol, methanol/ethanol, acetone, ionic liquids, supercritical $CO_2$, and ethyl acetate/ethyl lactate are recommended green solvents. Some of the less problematic solvents are toluene, heptane dimethylsulfoxide, tetrahydrofuran, acetonitrile, xylene, and dichloromethane, which could be used in as little as possible to minimize the toxicity associated with the solvent.

Various approaches have been proposed to achieve the synthesis of organometallic nanomaterials for specific purposes and preferable applications. Synthetic methods can be classified as conventional, microwave-assisted, electrochemical, sonochemical, and mechanochemical [10]. Conventional synthetic processes are either solvothermal or nonsolvothermal, which are basically temperature dependent [11]. Conventional approaches are commonly very slow, and the chemical reactions involved in the process take a long time to complete. Microwave-assisted approaches have been developed to control reaction rate issues; consequently, the process is very rapid [12]. In the microwave-assisted method, solvents are selected based on microwave absorption ability rather than the boiling point. Microwave-assisted approaches are suitable for synthesizing core-shell-type organometallic nanomaterials because the particle size can be precisely controlled. The reported chapters suggest that the linker type and irradiation time can potentially control the properties of synthesized materials [13–15].

Electrochemical synthetic techniques are suitable for ceaseless manufacturing of materials where the selected electrolyte is dissolved in the reaction mixture. Different types of electrochemical synthetic methods have been previously explored to fabricate porous metal nanoparticles of platinum and its alloys [16]. Apart from porous nanoparticles, other types of organometallic substructures, such as predesigned thin layers, might be carefully prepared by controlling the reaction conditions [17]. Organometallic octahedral crystals of 100–200 nm might be created by copper accumulation on the glass substrate by a galvanic replacement strategy [18]. Organometallic nanomaterials utilized for catalysis or drug delivery systems are manufactured by this method.

Rarely used, sonochemical synthesis applies very strong ultrasound [19, 20]. However, this method is faster and more environmentally friendly than the solvothermal method [21]. Nanomaterial substructures of 10–200 nm can be achieved by altering the reaction time that has been obtained [22]. A noninterpenetrated structure can be constructed by low sonication power, while the fabrication of an interpenetrated structure is achieved by higher sonication power. The average sonication power may lead to the formation of a hybrid nanomaterial composed of nonpenetrated and penetrated structures.

In the case of mechanochemical synthesis, the chemical transformation of expected products occurs due to mechanical energy. This approach is superior to others because it does not require any solvent and can be conducted at room temperature [23]. In addition to the aforementioned advantages, this method provides flexibility to utilize metallic precursors. In the case of mechanochemical synthetic methods, metal oxide salts are suitable as precursors since no solvents are required [24]. The rate of mechanochemical reaction can be increased by adding a small quantity of solvent, which is referred to as liquid-assisted grinding [25].

## 11.3 Applications of Organometallic Nanomaterials in Sustainable Green Nanotechnology

Since the rapid development of new organometallic nanomaterials, we have witnessed versatile and innovative applications in various fields. The scopes of applications not only cover conventional aspects, including capturing, extraction, adsorption, and separation but also expand to other fields, such as biomedicine, advanced devices, photonics, heterogeneous catalysis, and sustainable energy [26].

### 11.3.1 Capture, Sensing, and Separation

Heavy metals are harmful to the environment, humans, animals, and other organisms. The application of organometallic nanomaterials could be a promising approach to detect and assess the toxic metal ions present in polluted water and the atmosphere by electrochemical identification techniques [27]. This electrochemical detection strategy is expeditious and cost-effective, and finding its huge demand for perceiving the selective determination and measuring the toxic levels of heavy metal ions in polluted areas. Organometallic nanomaterials are suitable candidates for electrodes used in electrochemical detection methods for sensing metal ions such as cadmium, lead, mercury, and arsenic due to their 3D structure with controllable pore sizes and a greater specific area. Uranium, found in aqueous solutions in the form of $UO_2^{2+}$, is an extremely toxic substance released from a nuclear power plant, industrial projects, and mining activities [28]. Exposure to uranium and chronic ingestion of such chemical species can lead to a high risk of cancer, liver damage, internal irradiation, and overall chemical toxicity.

Qin's group synthesized a fluorescent organometallic nanomaterial named HNU-50, composed of pyromellitic acid (PMA), *N*-pyridin-4-ylpyridine-4-carboxamide (NPYC), and Zn(II), for the detection and adsorption of U(VI) from aqueous solution [29]. The structure of HNU-50 is illustrated in Fig. 11.1. The presence of carboxylate oxygen atoms, Lewis base active sites, and amide groups greatly and effectively enhanced the ability to capture U(VI) ions. The performance of HUN-50 is satisfactory even after recycling, which is a very important feature to ensure the sustainable application of such materials.

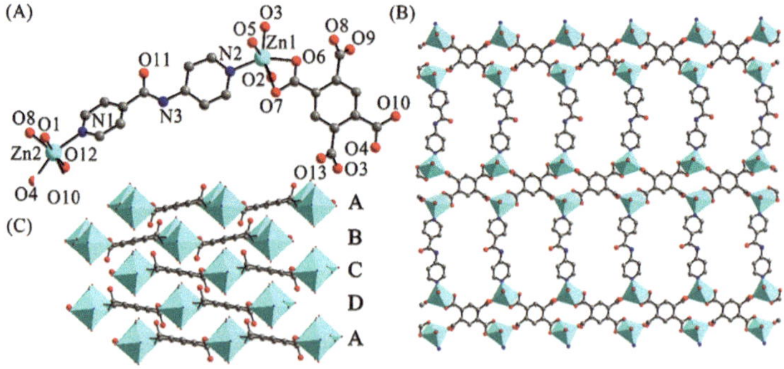


**Figure 11.1** (A), (B), and (C) represent the organization of the different components, monolayer frame, and three-dimensional substructure of HNU-50, respectively. The presence of oxygen, carbon, nitrogen, and zinc is represented by red, gray, dark blue, and light blue, respectively. Reprinted (adapted) with permission from Ref. 29, Copyright 2020, American Chemical Society.

Sulfur-derivatized porous organometallic nanomaterials, comprising benzene-1,3,5- tricarboxylic acid (BTC) and Cu(II), have been synthesized for the selective extraction of mercury, lead, nickel, cadmium, and zinc ions from polluted water samples [30]. The structural pattern of immobilized $(NH_4)_2MoS_4$ on the Cu-BTC organometallic nanomaterial and the selective extraction of mercury, lead, nickel, cadmium, and zinc ions from the wastewater sample are represented in Fig. 11.2. The immobilization of ammonium tetrathiomolybdate has improved the stability of sulfur-functionalized organometallic sorbents, having an excellent adsorption capability of 714.3–1000.0 mg per gram and the highest removal efficiency of 75.0–96.5%. The researchers involved with the work have claimed that the proposed sorbent shows recyclability and outstanding chemical stability that ensure the sustainable applications of this organometallic nanomaterial.

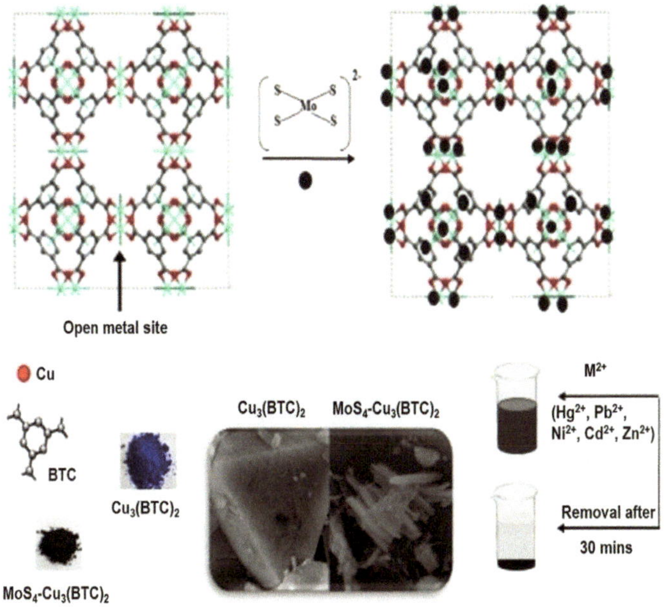


**Figure 11.2** Graphical representation of the structure of immobilized $(NH_4)_2MoS_4$ on Cu-BTC organometallic nanomaterial and the selective extraction of mercury, lead, nickel, cadmium, and zinc ions from a wastewater sample. Reprinted (adapted) with permission from Ref. 30, Copyright 2021, Elsevier.

Ferrous organometallic nanoparticle-based chemosensors with novel features and advanced functionalities have been synthesized by uncomplicated and cost-effective methods for detecting mercury ($Hg^{2+}$) present in biological and environmental samples [31]. The authors of the work have declared that they can overcome the challenges concerning $Hg^{2+}$ determination. The detection and removal processes of the proposed sensor are easy, fast, economical, sensitive, and highly selective. In addition, the sensor is easy to carry and operate and is suitable for analyzing water samples from different sources, such as minerals, tap, sea, river, and wastewater, and biological samples, including blood serum and urine. Wang and coworkers have invented ultrahigh sulfur-containing 2D organometallic nanosheets potentially suitable for mercury adsorption, with appreciable productivity, selectivity, and regeneration ability [32]. The mercury capturing mechanism and the regeneration process are shown in Fig. 11.3.

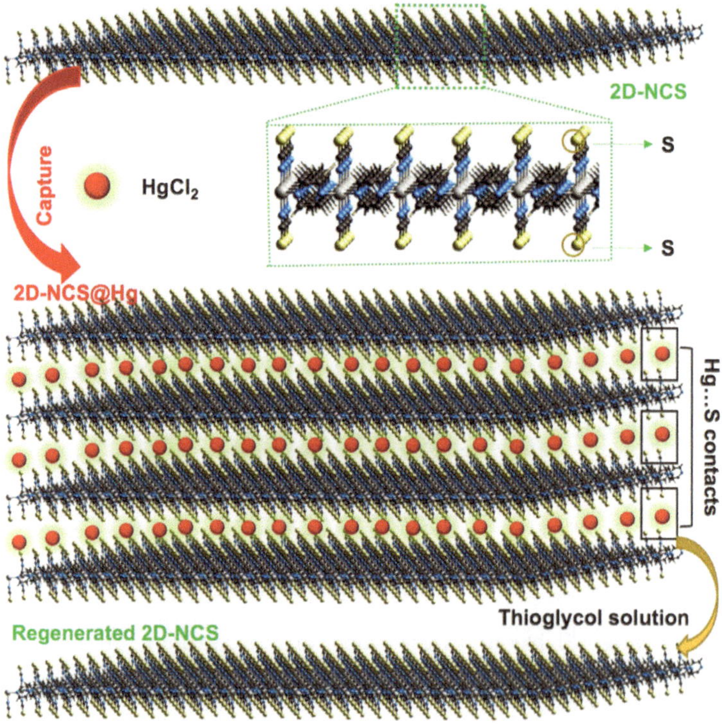


**Figure 11.3** Symbolic and simplified diagram for metal adsorption of two-dimensional organometallic nanosheets. Reprinted (adapted) with permission from Ref. 32, Copyright 2020, Elsevier.

The removal of arsenic from contaminated water can be accomplished by employing manganese dioxide ($MnO_2$)-embedded carbon nanotube ($MnO_2$@CNT)-based organometallic adsorbents where carbon fiber (CF) is used as a substrate [33]. A novel adsorbent has been synthesized for the extraction of heavy metals and organic dyestuff [34]. Chitosan aerogels show better water stability as BC nanofibers have been incorporated through a physical mixing process rather than using a chemical crosslinking agent, which is toxic in nature and potentially hazardous to the environment. Figure 11.4 shows the extraction mechanism, while Fig. 11.5 shows the fabrication process and digital photographs of different aerogels. Polymeric hydrogels have also been investigated for dye and metal removal from wastewater [35, 36].

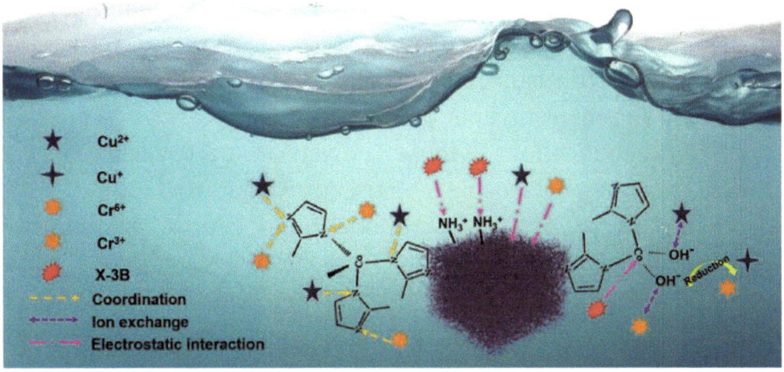


**Figure 11.4** Symbolic and simplified representation of the metal and dye removal mechanism of an organometallic aerogel nanocomposite (ZIF-67). Reprinted (adapted) with permission from Ref. 34, Copyright 2020, Elsevier.

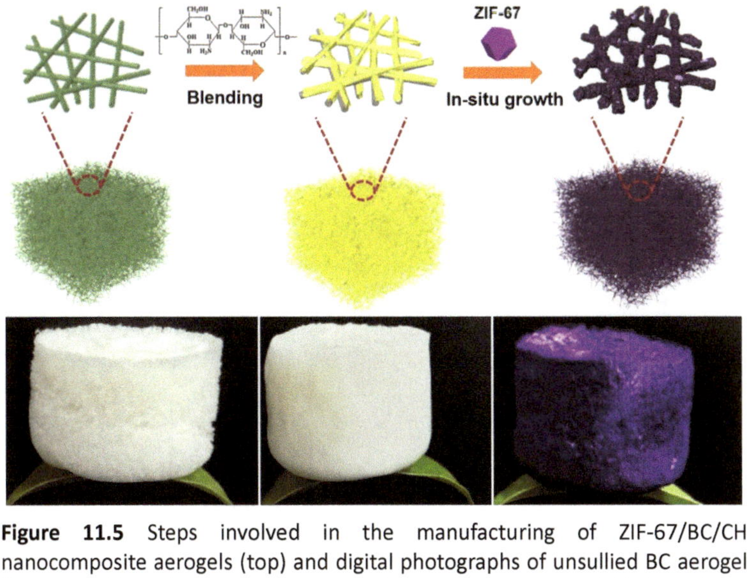


**Figure 11.5** Steps involved in the manufacturing of ZIF-67/BC/CH nanocomposite aerogels (top) and digital photographs of unsullied BC aerogel (bottom left), BC/CH dual nanocomposite aerogel (bottom middle), and ZIF-67/BC/CH nanocomposite aerogel (bottom right). Reprinted (adapted) with permission from Ref. 34, Copyright 2020, Elsevier.

## 11.3.2 Advanced Catalysis

Organometallic nanomaterials can efficiently be utilized to synthesize catalysts with diverse functionalities, including catalysts capable of outstanding intrinsic reactivity and multifunctional catalysts for various organic transformations and as supporting materials for prevalent catalysts [37]. Organometallic nanomaterial-based enzyme-mimetic catalysts (Enz-Cats) have attracted researchers due to their better catalytic performance and selectivity in various physiological environments than natural enzymes. Some potential biomedical applications of these catalysts are bacterial disinfection, tumor therapies, biosensors, and tissue regeneration [38].

Fu et al. used a postsynthetic modification to produce a phosphorus–nitrogen-containing precursor to prepare highly efficient oxygen reduction reaction electrocatalysts [39]. As illustrated in Fig. 11.6, a sturdy organometallic nanomaterial containing nitrogen was chosen as the starting precursor. The nitrogen–phosphorus codoped carbon precursor with a uniform phosphorus distribution in the substructure was obtained by utilizing an appropriate phosphorus source, glyphosine, to react with the amine groups present in the organometallic substructure. Afterward, the nitrogen–phosphorus–carbons were obtained by a single-step pyrolysis method. These newly fabricated organometallic nanomaterials demonstrate excellent catalytic activities similar to those of commercially available Pt catalysts. In addition, the catalyst shows outstanding methanol forbearance and exceptional longevity.

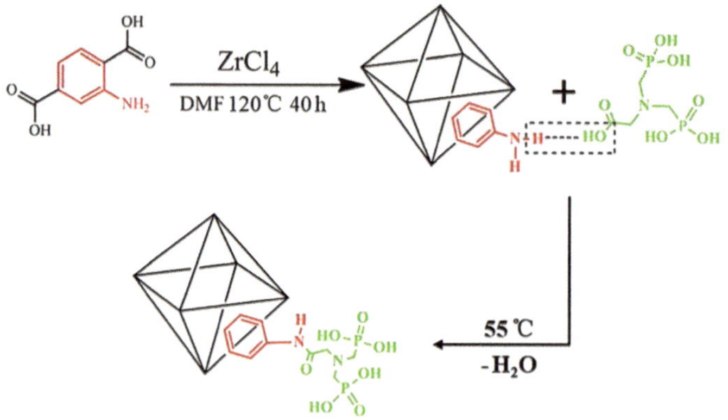


**Figure 11.6** Preparation of the nitrogen–phosphorus-rich carbon precursor. Reprinted (adapted) with permission from Ref. 39, Copyright © 2016, Wiley-VCH Verlag GmbH & Co.

Organometallic nanomaterials have great potential in the field of electrocatalysis [40]. These materials may find a very effective application as highly active electrocatalysts to reduce carbon dioxide from carbonaceous fuels and nitrogen reduction during ammonia production, which are two important energy-related reactions.

### 11.3.3 Biomedicine

Organometallic nanostructured materials are potentially suitable for biomedical applications due to their distinctive properties, including high porosity, large surface area, and biodegradability [41]. The synthetic modification ability of organometallic substructures with high loading capacities and biocompatibility makes them satisfactory candidates for drug delivery systems. Orellana-Tavira's group has reported the use of CAU-7 ([Bi(BTB)], with BTB=1,3,5-benzenetrisbenzoate and CAU is the abbreviation of Christian–Albrechts–Universität, a drug delivery system), which is a biocompatible bismuth-based organometallic substructure suitable for drug delivery in cancer treatment [42]. Wang et al. reported the fabrication of organometallic nanomaterials with special rod-on-octahedron morphology for a double-drug delivery method with high precision and accuracy [43]. Figure 11.7 illustrates the drug delivery system.

Furthermore, organometallic nanoparticles have been synthesized to adsorb protein Alexa Fluor 647-labeled donkey anti-rabbit IgG [44]. These unique solid substrates for protein adsorption may find potential application for medicine, food production, pharmaceutical processing, and biotechnology. Fu's group proposed organometallic nanofibrous materials (Fig. 11.8) with enhanced protein adsorption capacity compared to a material composed of polyethylene terephthalate gauze fiber. The research group has demonstrated that the reported material offers biologically well-matched characteristics for managing the working principle of enzymes, which is essential for sustainable nanotechnology applications. Organometallic nanomaterials have found successful applications in immunoassays and biomarker detection. By applying a $TiO_2$-based organometallic nanomaterial as the promoter, an ovarian cancer biomarker has been detected efficiently and accurately [45]. Secondary antibody-coated organometallic nanomaterials may serve as an effective candidate to ease fluorescent indication expansion for immunoassessments of cardiac troponin I with enhanced sensitivity and a wide detection range [46].

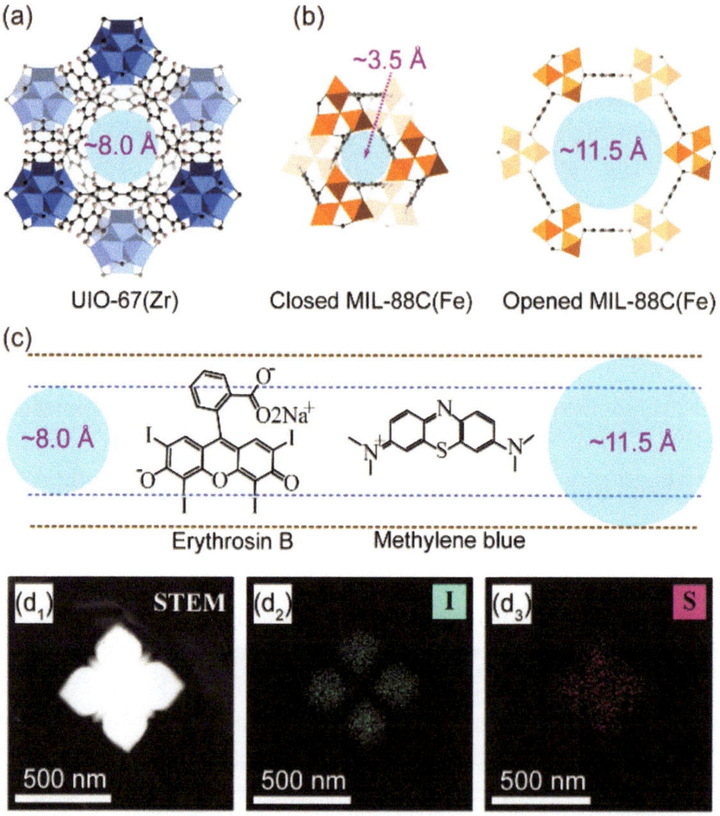


**Figure 11.7** Graphical representation of (a) UIO-67 and (b) MIL-88C. (c) The location of the atoms, groups or ions relative to one another in erythrosin B and methylene blue and their pore size contrast with UIO-67 and MIL-88C. (d1-d3) STEM-EDX analysis of MIL-88C-on-UIO-67. Reprinted (adapted) with permission from Ref. 43, Copyright 2020, John Wiley and Sons.

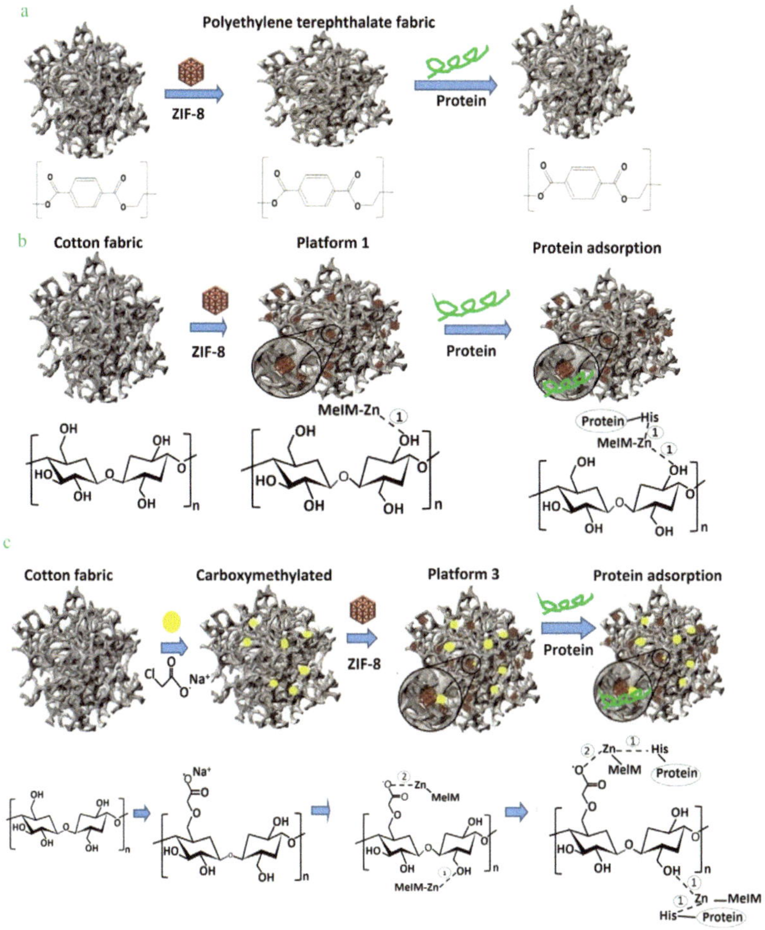


**Figure 11.8** (a) Protein extraction failure by untreated surface. (b) Effective extraction by a fibrous surface modified by ZIF-8 organometallic nanomaterials. (c) Carboxymethylation of the fibrous surface, extraction of ZIF-8, and protein adsorption. The dotted connection (1) shows chelation and dual electrostatic force of attraction. Reprinted (adapted) with permission from Ref. 44, Copyright 2019, American Chemical Society.

## 11.3.4 Energy and Light

The utilization of organometallic nanomaterials and their derivatives for imperishable energy harnessing is an advanced and innovative theme. Organometallic nanomaterials have shown immense prospects for clean and renewable energy production through sustainable technologies such as adsorption technologies for heat conversion and water harvesting [47]. Pristine organometallic nanomaterials, their derivatives, and composites provide numerous platforms for energy technology development, such as chemical hydrogen storage, physical hydrogen and methane storage, solar energy conversion, and electrochemical conversion and storage [48]. Cobalt-based organometallic nanomaterials with structural versatility, tunable pore sizes, large surface areas, cobalt central nodes, and unique electrochemical properties could be utilized for electrochemical energy storage [49]. In terms of the sustainable energy sector, organometallic nanomaterial-based nanostructures are applicable for Li-ion battery construction and modification [50]. Wang et al. proposed a straightforward approach to develop self-intertwined carbonaceous organometallic nanomaterials to overcome censorious obstacles associated with conventional approaches (Fig. 11.9). The formation of the aforementioned substructure is due to the disintegration of the sacrificial organometallic nanomaterials into nanoparticles. The porous structure offers high performance for Li-ion batteries. Iron(II)-containing organometallic nanocrystals have been built for fuel cells [51].

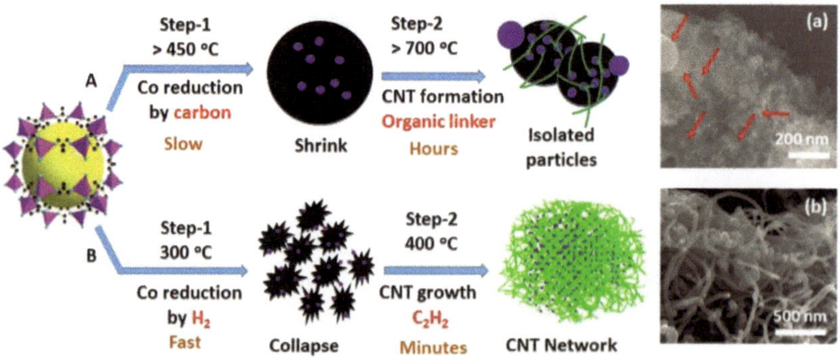


**Figure 11.9** Graphical representation of different steps associated with the fabrication of carbonaceous organometallic nanomaterials by conventional (method A) and advanced strategies (method B). Purple spheres represent cobalt nanoparticles. Scanning electron micrographs of material (a) achieved through the conventional method [52]; red arrows indicate agglomeration of metallic nanoparticles; (b) ZCNT-400 achieved through the advanced method. Reprinted (adapted) from Ref. 50, Copyright 2018, American Chemical Society.

Pure or hybridized organometallic nanomaterials exhibiting photonic characteristics have found applications in designing intrinsic, extrinsic, and intrinsic–extrinsic hybrid luminescent units displaying synergistic effects. These materials are effective for the miniaturization of lasers. In addition, photoabsorbing polymer-coated biocompatible organometallic nanoscale composites are utilized in lighting, nonlinear optics, lasing, and phototherapy in cancer treatment [53]. Multilayered organometallic thin films are promising for high-performance electrochemical optical sensors [54]. Furthermore, the pores within organometallic substructures can also be applicable to envelope luminescent guest species. They can be utilized to fabricate organometallic nanomaterials with an extended luminescent character that is not available in commonly found organic and inorganic luminescent materials. Organometallic nanomaterials showing enhanced luminescent properties have been manufactured for applications in white-light-emitting diodes, sensing, and photocatalysis [55].

## 11.4 Future Outlook and Challenges

The fabrication approaches of organometallic nanomaterials have already been transformed from small-batch manufacturing to large-scale industrial manufacturing. A good number of industrial manufacturers have already synthesized these materials for commercial applications such as catalysis, the energy sector, and the environment. The evolution of application-oriented organometallic nanomaterials is generally centered on selecting and screening substructure designs through computational approaches. In addition to the design and development of such nanomaterials, the defiance of production methods is fairly essential to consider. Safety risks and negative environmental effects might be found to be unimportant in lab production but must be taken care of seriously for mass manufacturing. Moreover, the functionality, biocompatibility, price, biodegradability, facile and environmentally friendly manufacturing of organometallic nanomaterials should be considered for the commercial market and consumers. At present and in the near future, it would be a great concern of manufacturers and consumers to design and implement clean and sustainable techniques for production. It is essential to evaluate the safety of these nanomaterials for health and the environment [56, 57]. The structural topology, metal ions, organic ligands, size, shape, and pore size of organometallic materials may contribute to toxicity [58]. The amalgamation of green chemistry and sustainable practice in industrial production may lead to the development and implementation of eco-friendly synthetic methods that can offer rapid growth in practical applications of organometallic nanomaterials in green nanotechnology.

## 11.5 Conclusion

There are a large number of original research papers, letters, communications, editorials, book chapters, and review articles on the manufacturing, characterization, applications, and improvement of organometallic materials in general. However, to the best of our knowledge, a dedicated chapter, especially on organometallic nanomaterials and their sustainable application in green nanotechnology, is absent to date. The scarcity of such chapters and the importance of sustainable green nanotechnology have motivated us to write this chapter exclusively allocated for organometallic nanomaterial synthesis and their prospective applications in sustainable green nanotechnology. Traditional pathways are well-liked tools for the synthesis of organometallic materials. However, traditional methods are not ecologically suitable due to the association of toxic and hazardous components and manufacturing processes. The evolution and introduction of biocompatible organic ligands, earth-abundant benign metal ions, and green synthetic approaches are promising and would be effective in replacing conventional and popular but environmentally hazardous manufacturing processes. Although a substantial number of research efforts have been put together in developing organometallic materials, there is still a great deal of scope and demand for sustainable synthetic procedures and commercially available green nanotechnology products for general consumers. Production cost is another crucial factor in addition to the hazardous components of organometallic nanomaterials. The selection of biobased ligands like 2,5-furandicarboxylic acid and lactic acid may result in optimum production costs since such ligands are cheaper and easily extractable. Exploring new research, a better understanding of health and environmental risks, and the effective implementation of green and sustainable processes involved in organometallic nanomaterial fabrication would ensure sustainable green nanotechnology applications.

## Acknowledgments

Abu Bin Imran gratefully acknowledges the support of the Capacity Utilization Program under Special Allocation for Science and Technology (BS-182 and PHY'S-467) from the Ministry of Science and Technology, the People's Republic of Bangladesh. The author is also thankful to the Committee for Advanced Studies and Research (CASR) in BUET for funding.

## Conflicts of Interest

The authors declare that there are no conflicts of interest.

## References

1. Kumar, S., Jain, S., Nehra, M., Dilbaghi, N., Marrazza, G., and Kim, K. H. (2020). Green synthesis of metal–organic frameworks: A state-of-the-art review of potential environmental and medical applications. *Coordination Chemistry Reviews*, **420**, 213407.
2. Li, P., Cheng, F. F., Xiong, W. W., and Zhang, Q. (2018). New synthetic strategies to prepare metal–organic frameworks. *Inorganic Chemistry Frontiers*, **5**(11), 2693–2708.
3. Yu, J., Li, X., and Deria, P. (2018). Light-harvesting in porous crystalline compositions: Where we stand toward robust metal–organic frameworks. *ACS Sustainable Chemistry* and *Engineering*, **7**(2), 1841–1854.
4. Li, L., He, J., Wang, Y., Lv, X., Gu, X., Dai, P., Liu, D., and Zhao, X. (2019). Metal–organic frameworks: a promising platform for constructing nonnoble electrocatalysts for the oxygen-reduction reaction. *Journal of Materials Chemistry A*, **7**(5), 1964–1988.
5. Wang, C., An, B., and Lin, W. (2018). Metal–organic frameworks in solid–gas phase catalysis. *ACS Catalysis*, **9**(1), 130–146.
6. Engel, E. R., and Scott, J. L. (2020). Advances in the green chemistry of coordination polymer materials. *Green Chemistry*, **22**(12), 3693–3715.

7. Asakawa, M., Shrotri, A., Kobayashi, H., and Fukuoka, A. (2019). Solvent basicity controlled deformylation for the formation of furfural from glucose and fructose. *Green Chemistry*, **21**(22), 6146–6153.
8. Prat, D., Wells, A., Hayler, J., Sneddon, H., McElroy, C. R., Abou-Shehada, S., and Dunn, P. J. (2015). CHEM21 selection guide of classical-and less classical-solvents. *Green Chemistry*, **18**(1), 288–296.
9. Byrne, F. P., Jin, S., Paggiola, G., Petchey, T. H., Clark, J. H., Farmer, T. J., Hunt, A.J., McElroy, C.R. and Sherwood, J. (2016). Tools and techniques for solvent selection: Green solvent selection guides. *Sustainable Chemical Processes*, **4**(1), 1–24.
10. Pandey, A., Dhas, N., Deshmukh, P., Caro, C., Patil, P., García-Martín, M. L., Padya, B., Nikam, A., Mehta, T., and Mutalik, S. (2020). Heterogeneous surface architectured metal-organic frameworks for cancer therapy, imaging, and biosensing: A state-of-the-art review. *Coordination Chemistry Reviews*, **409**, 213212.
11. Butova, V. V. E., Soldatov, M. A., Guda, A. A., Lomachenko, K. A., and Lamberti, C. (2016). Metal-organic frameworks: structure, properties, methods of synthesis and characterization. *Russian Chemical Reviews*, **85**(3), 280.
12. Huang, X. C., Lin, Y. Y., Zhang, J. P., and Chen, X. M. (2006). Ligand-directed strategy for zeolite-type metal–organic frameworks: Zinc (II) imidazolates with unusual zeolitic topologies. *Angewandte Chemie International Edition*, **45**(10), 1557–1559.
13. Li, J., Cheng, S., Zhao, Q., Long, P., and Dong, J. (2009). Synthesis and hydrogen-storage behavior of metal–organic framework MOF-5. *International Journal of Hydrogen Energy*, **34**(3), 1377–1382.
14. Jhung, S. H., Lee, J. H., and Chang, J. S. (2005). Microwave synthesis of a nanoporous hybrid material, chromium trimesate. *Bulletin of the Korean Chemical Society*, **26**(6), 880–881.
15. Seo, Y. K., Hundal, G., Jang, I. T., Hwang, Y. K., Jun, C. H., and Chang, J. S. (2009). Microwave synthesis of hybrid inorganic–organic materials including porous $Cu_3(BTC)_2$ from Cu(II)-trimesate mixture. *Microporous and Mesoporous Materials*, **119**(1–3), 331–337.
16. Schlesinger, M., Schulze, S., Hietschold, M., and Mehring, M. (2010). Evaluation of synthetic methods for microporous metal–organic frameworks exemplified by the competitive formation of $Cu_2(btc)_3(H_2O)_3$. and $Cu_2$ (btc)(OH)($H_2O$).. *Microporous and Mesoporous Materials*, **132**(1–2), 121–127.

17. Ameloot, R., Stappers, L., Fransaer, J., Alaerts, L., Sels, B. F., and De Vos, D. E. (2009). Patterned growth of metal-organic framework coatings by electrochemical synthesis. *Chemistry of Materials*, **21**(13), 2580–2582.

18. Ameloot, R., Pandey, L., Van der Auweraer, M., Alaerts, L., Sels, B. F., and De Vos, D. E. (2010). Patterned film growth of metal–organic frameworks based on galvanic displacement. *Chemical Communications*, **46**(21), 3735–3737.

19. Qiu, L. G., Li, Z. Q., Wu, Y., Wang, W., Xu, T., and Jiang, X. (2008). Facile synthesis of nanocrystals of a microporous metal–organic framework by an ultrasonic method and selective sensing of organoamines. *Chemical Communications*, (31), 3642–3644.

20. Son, W. J., Kim, J., Kim, J., and Ahn, W. S. (2008). Sonochemical synthesis of MOF-5. *Chemical Communications*, (47), 6336–6338.

21. Li, Z. Q., Qiu, L. G., Xu, T., Wu, Y., Wang, W., Wu, Z. Y., and Jiang, X. (2009). Ultrasonic synthesis of the microporous metal–organic framework $Cu_3(BTC)_2$ at ambient temperature and pressure: An efficient and environmentally friendly method. *Materials Letters*, **63**(1), 78–80.

22. Khan, N. A., and Jhung, S. H. (2009). Facile syntheses of metal-organic framework $Cu_3(BTC)_2(H_2O)_3$ under ultrasound. *Bulletin of the Korean Chemical Society*, **30**(12), 2921–2926.

23. Garay, A. L., Pichon, A., and James, S. L. (2007). Solvent-free synthesis of metal complexes. *Chemical Society Reviews*, **36**(6), 846–855.

24. Yuan, W., Friščić, T., Apperley, D., and James, S. L. (2010). High reactivity of metal–organic frameworks under grinding conditions: parallels with organic molecular materials. *Angewandte Chemie International Edition*, **49**(23), 3916–3919.

25. Friščić, T., Reid, D. G., Halasz, I., Stein, R. S., Dinnebier, R. E., and Duer, M. J. (2010). Ion-and liquid-assisted grinding: Improved mechanochemical synthesis of metal–organic frameworks reveals salt inclusion and anion templating. *Angewandte Chemie*, **122**(4), 724–727.

26. Chowdhury, A.-N., Shapter, J., and Imran, A. B. (2015). *Innovations in Nanomaterials.* Nova Science Publishers, Inc., NY, USA, **1**, pp. 1-422. ISBN: 978-1-63483-548-0.

27. Devaraj, M., Yesudass, S., Rajendran, S., and Ponce, L.C. (2021). Metal organic framework based nanomaterials for electrochemical sensing of toxic heavy metal ions: Progress and their prospects. *Journal of the Electrochemical Society*, 168, 037513.

28. Li, J., Wang, X., Zhao, G., Chen, C., Chai, Z., Alsaedi, A., Hayat, T., and Wang, X. (2018). Metal–organic framework-based materials: superior adsorbents for the capture of toxic and radioactive metal ions. *Chemical Society Reviews*, **47**(7), 2322–2356.

29. Qin, X., Yang, W., Yang, Y., Gu, D., Guo, D., and Pan, Q. (2020). A zinc metal–organic framework for concurrent adsorption and detection of uranium. *Inorganic Chemistry*, **59**(14), 9857–9865.

30. Yazdi, M. N., Dadfarnia, S., and Shabani, A. M. H. (2021). Synthesis of stable S-functionalized metal-organic framework using $MoS_4^{2-}$ and its application for selective and efficient removal of toxic heavy metal ions in wastewater treatment. *Journal of Environmental Chemical Engineering*, **9**(1), 104696.

31. Basaleh, A. S., and Sheta, S. M. (2020). Novel advanced nanomaterial based on ferrous metal–organic framework and its application as chemosensors for mercury in environmental and biological samples. *Analytical and Bioanalytical Chemistry*, **412**(13), 3153–3165.

32. Wang, C., He, C., Luo, Y. H., Su, S., Wang, J. Y., Hong, D. L., He, X.T., Chen, C., and Sun, B. W. (2020). Efficient mercury chloride capture by ultrathin 2D metal-organic framework nanosheets. *Chemical Engineering Journal*, **379**, 122337.

33. Sridhar, V., Lee, I., Jung, K. H., and Park, H. (2020). Metal organic framework derived $MnO_2$-carbon nanotubes for efficient oxygen reduction reaction and arsenic removal from contaminated water. *Nanomaterials*, **10**(9), 1895.

34. Li, D., Tian, X., Wang, Z., Guan, Z., Li, X., Qiao, H., Ke, H., Luo, L., and Wei, Q. (2020). Multifunctional adsorbent based on metal-organic framework modified bacterial cellulose/chitosan composite aerogel for high efficient removal of heavy metal ion and organic pollutant. *Chemical Engineering Journal*, **383**, 123127.

35. Chowdhury, N., Solaiman., Roy, C. K., Firoz, S. H., Foyez, T., Imran, A. B. (2020). Role of ionic moieties in hydrogel networks to remove heavy metal ions from water, *ACS Omega*, **6**(1), 836–844.

36. Rahman, A., Solaiman., Foyez, T., Susan, M. A. B. H., and Imran, A. B. (2020). Self-healable and conductive double-network hydrogels with bioactive properties. *Macromolecular Chemistry and Physics*, 221, 202000207.

37. Tan, C., Liu, G., Li, H., Cui, Y., and Liu, Y. (2020). Ultrathin two-dimensional metal–organic framework nanosheets: An emerging class of catalytic nanomaterials. *Dalton Transactions*, **49**(32), 11073–11084.

38. Ma, L., Jiang, F., Fan, X., Wang, L., He, C., Zhou, M., Li, S., Luo, H., Cheng, C., and Qiu, L. (2020). Metal–organic-framework-engineered enzyme-mimetic catalysts. *Advanced Materials*, **32**(49), 2003065.

39. Fu, Y. A., Huang, Y., Xiang, Z., Liu, G., and Cao, D. (2016). Phosphorous–nitrogen-codoped carbon materials derived from metal–organic frameworks as efficient electrocatalysts for oxygen reduction reactions. *European Journal of Inorganic Chemistry*, **2016**(13–14), 2100–2105.

40. Singh, C., Mukhopadhyay, S., and Hod, I. (2021). Metal–organic framework derived nanomaterials for electrocatalysis: Recent developments for $CO_2$ and $N_2$ reduction. *Nano Convergence*, **8**(1), 1–10.

41. Zhang, S., Pei, X., Gao, H., Chen, S., and Wang, J. (2020). Metal-organic framework-based nanomaterials for biomedical applications. *Chinese Chemical Letters*, **31**(5), 1060–1070.

42. Orellana-Tavra, C., Köppen, M., Li, A., Stock, N., and Fairen-Jimenez, D. (2020). Biocompatible, crystalline, and amorphous bismuth-based metal–organic frameworks for drug delivery. *ACS Applied Materials and Interfaces*, **12**(5), 5633–5641.

43. Wang, X. G., Xu, L., Li, M. J., and Zhang, X. Z. (2020). Construction of flexible-on-rigid hybrid-phase metal–organic frameworks for controllable multi-drug delivery. *Angewandte Chemie International Edition*, **59**(41), 18078–18086.

44. Fu, H., Ou, P., Zhu, J., Song, P., Yang, J., and Wu, Y. (2019). Enhanced protein adsorption in fibrous substrates treated with zeolitic imidazolate framework-8 (ZIF-8) nanoparticles. *ACS Applied Nano Materials*, **2**(12), 7626–7636.

45. Zhang, S., Zheng, H., Chen, Y., Yi, H., Dai, H., Hong, Z., and Lin, Y. (2019). Electrochemiluminescence resonance energy transfer between Ru (bpy) 32+ and CdZnSe@ ZnSe quantum dots for ovarian cancer biomarker detection. *ACS Applied Nano Materials*, **2**(11), 7061–7066.

46. Wang, X., Wang, X., Han, Y., Li, H., Kang, Q., Wang, P., and Zhou, F. (2019). Immunoassay for cardiac troponin I with fluorescent signal amplification by hydrolyzed coumarin released from a metal–organic framework. *ACS Applied Nano Materials*, **2**(11), 7170–7177.

47. Gordeeva, L. G., Tu, Y., Pan, Q., Palash, M. L., Saha, B. B., Aristov, Y. I., and Wang, R. (2021). Metal-organic frameworks for energy conversion and water harvesting: A bridge between thermal engineering and material science. *Nano Energy*, 105946.

48. Li, X., Yang, X., Xue, H., Pang, H., and Xu, Q. (2020). Metal–organic frameworks as a platform for clean energy applications. *EnergyChem*, **2**(2), 100027.

49. Li, S., Lin, J., Xiong, W., Guo, X., Wu, D., Zhang, Q., Zhu, Q. L., and Zhang, L. (2021). Design principles and direct applications of cobalt-based metal-organic frameworks for electrochemical energy storage. *Coordination Chemistry Reviews*, **438**, 213872.

50. Wang, X., Yin, H., Sheng, G., Wang, W., Zhang, X., and Lai, Z. (2018). Fabrication of self-entangled 3D carbon nanotube networks from metal–organic frameworks for Li-ion batteries. *ACS Applied Nano Materials*, **1**(12), 7075–7082.

51. Bunzen, H., Javed, A., Klawinski, D., Lamp, A., Grzywa, M., Kalytta-Mewes, A., Tiemann, M., von Nidda, H. A. K., Wagner, T., and Volkmer, D. (2018). Anisotropic water-mediated proton conductivity in large iron (II) metal–organic framework single crystals for proton-exchange membrane fuel cells. *ACS Applied Nano Materials*, **2**(1), 291–298.

52. Zhang, W., Jiang, X., Wang, X., Kaneti, Y. V., Chen, Y., Liu, J., Jiang, J. S., Yamauchi, Y., and Hu, M. (2017). Spontaneous weaving of graphitic carbon networks synthesized by pyrolysis of ZIF-67 crystals. *Angewandte Chemie International Edition*, **56**(29), 8435–8440.

53. Cong, C., and Ma, H. (2021). Photonic metal-organic frameworks. *Advanced Optical Materials*, 2100733.

54. Ikigaki, K., Okada, K., and Takahashi, M. (2021). Epitaxial growth of multilayered metal–organic framework thin films for electronic and photonic applications. *ACS Applied Nano Materials*, **4**(4), 3467–3475.

55. Shu, Y., Ye, Q., Dai, T., Xu, Q., and Hu, X. (2021). Encapsulation of luminescent guests to construct luminescent metal–organic frameworks for chemical sensing. *ACS Sensors*, **6**(3), 641–658.

56. Goswami, L., Kim, K. H., Deep, A., Das, P., Bhattacharya, S. S., Kumar, S., and Adelodun, A. A. (2017). Engineered nano particles: nature, behavior, and effect on the environment. *Journal of Environmental Management*, **196**, 297–315.

57. Tamames-Tabar, C., Cunha, D., Imbuluzqueta, E., Ragon, F., Serre, C., Blanco-Prieto, M. J., and Horcajada, P. (2014). Cytotoxicity of nanoscaled metal–organic frameworks. *Journal of Materials Chemistry B*, **2**(3), 262–271.

58. Kumar, P., Anand, B., Tsang, Y. F., Kim, K. H., Khullar, S., and Wang, B. (2019). Regeneration, degradation, and toxicity effect of MOFs: Opportunities and challenges. *Environmental Research*, **176**, 108488.

## Chapter 12

# Green Nanomaterials Revolution in Cosmetic Products and Skin Treatment

**Nahid Rehman* and Anjana Pandey***
*Department of Biotechnology, Motilal Nehru National Institute of Technology, Allahabad, Prayagraj, Uttar Pradesh, India*
anjanap@mnnit.ac.in

This chapter would emphasize using green nanomaterials as an ecofriendly safe alternative to chemical-based nanomaterial used for cosmetics and skin treatment. The cosmetic industry uses nanoparticles (NPs) to achieve long-lasting effects with increased consistency. Nanotechnology-based cosmetics have certain distinct advantages over micro-scale cosmetics. These NPs have a large surface area, which allows for more effective substance delivery through the skin. The effective diffusion into the skin for increased delivery of the product's elements, including new color aspects (e.g., lipsticks, blushes, and nail polishes), pellucidity (e.g., sunscreens), and long-term benefits could all be the advantages of using nanomaterials in

**The authors contributed equally to this work.*

*Green Nanoarchitectonics*
Edited by Kaushik Pal

ISBN 978-981-4968-50-8 (Hardcover), 978-1-003-31860-6 (eBook)
www.jennystanford.com

cosmetics (e.g., in makeup). The widespread cosmetic usage with less negative impact demand outpoured by the consumers has directed to the progress of inventions by switching to green approach in cosmetics not compromising with respect to performance, retentions, striking appearance, and more notably safety. When using NPs, the cosmetic industry's ultimate goal is to deliver the exact amount needs to the desired regions of the body while maintaining longer shelf life. In current epochs, green nanotechnology is a relative and multidisciplinary field that has developed as a rapidly emerging research area. Green nanomaterials are reported to possess significant advantages: low cost, high efficiency, unbiased environmental impact, as well as greater stability. Green Nanotechnology delivers inclusive information about green nanomaterials, their types, and methods for generation, characterization, and their properties. In a work by Radmard et al. (2021), Arbutin niosome (Arbusome); an ecological and green preparation revealed the skin permeation evaluations exhibited arbutin deposits in a superior amount in the layers of skin, as well as cytotoxicity of augmented Arbusome in lipid nanotechnology for handling and control of hyperpigmentation complaint, was not revealed in Dimethyl Tetrazole (MTT) assay. Thus indicating a positive response for welcoming green nanomaterial in cosmeceuticals.

## 12.1 Introduction

Nanotechnology had been acquainted for over four decades with the arena of cosmeceuticals, skin preparations, and health products. The term "cosmetics" was invented by Raymond Reed, one of the establishing members of the Cosmetic Chemists US Society, in 1961 [17]. Products that are acknowledged to promote the beautification of the skin, perform deep cleansing, and boost the appearance of the skin are regarded as Cosmetics. The Egyptians were the first to employ cosmetics circa 4000 BC, and later the Americans, Romans, Japanese, Chinese, and Greeks followed suit. Numerous diagnostic modalities and treatments for skin disease have been transformed as a result of the implementation of this unique technology. Cosmetics firms use nanoscale variations of chemicals to give stronger ultraviolet (UV) protection, better skin penetration, longer-lasting benefits, improved color, finish quality, etc. [8]. This widespread usage of nanoscale-based constituents in cosmetics owes to the fact that NPs attain sturdier properties that differ from larger particles. Color shade, transparency, solubility, and chemical reactivity are just a few of the features that have been altered, making nanomaterials appealing to the cosmetics and personal care industry [7].

Nanocosmeceuticals have several promising applications over conventional cosmetics, including improved stability, sensorial properties, and high entrapment efficiency, as well as controlled drug release from carriers due to a variety of factors, including preparation techniques, drug composition, extracts, and polymers used, as well as the biochemical or physical association among diverse components [11].

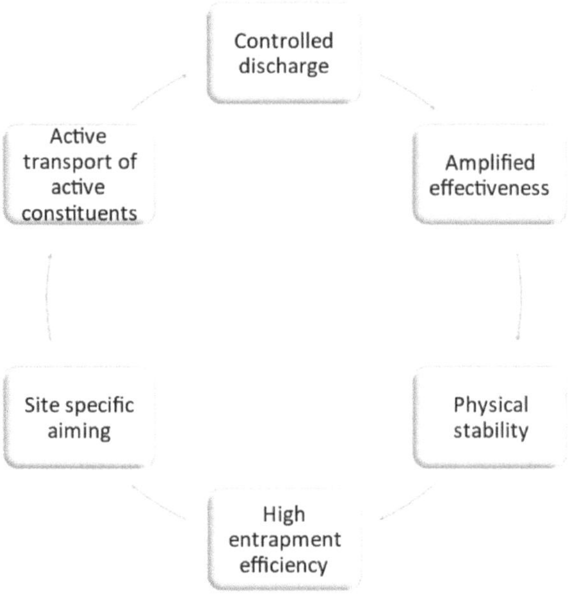


**Figure 12.1** Pictorial presentation of constructive characteristics of nanocosmeceuticals.

Nanocosmeceuticals are included in a variety of products meant for hair care, including shampoos from Nirvel and Origem. Allure Parfum uses nano-based materials in perfumes to make them stay longer [1]. Nanoparticles boost the efficacy of sunscreens and other skincare formulas by increasing protection against UV rays. Because the active chemicals are dynamically delivered to the skin due to the insignificant diameters of NPs, which enhances the surface range of cosmetics, thus desirable results are attained. Apart from frequently being used in hair serums, conditioners, and shampoos for hair repairment, NPs are widely incorporated in skin-lightening, nourishing, and anti-wrinkle emulsions [21].

Green nanotechnology has been reinvented with two goals in mind: (i) creating nanomaterials and products that do not possess a threat to the atmosphere or human well-being, and (ii) creating nano-based goods that are a solution to existing environmental concerns. It practices prevailing green chemistry and engineering principles for nanomaterials and nano-products creation avoiding the use of harmful substances, at ambient temperature with much less energy and renewable inputs whenever possible, and with biodegradable ideas at all design and engineering phases.

Green nanotechnology also means exploiting nanotechnology to make present production processes more environmentally friendly for non-nano materials and products. To help society and contribute to the economy it is essential to maintain hygiene and a healthy environment - United Nations Environment Programme.

## 12.2 Need for Green Nanomaterials in Cosmeceuticals

To begin with, the frequently reported ingredient in sunscreen is titanium dioxide ($TiO_2$) and zinc oxide (ZnO) NPs as they possess the ability to absorb UV light. Silver NPs (AgNPs) are used as preservatives in toothpaste, burn healing balm and shampoos due to their intrinsic antibacterial properties. Gold NPs (GNPs) based face masks have been created by Oro Gold Cosmetics as well. The GNPs present in the face mask, according to the business, have anti-inflammatory and antioxidant characteristics, as well as the ability to accelerate tissue renewal, restoration of skin suppleness, and minimize indications of stress besides aging. Regardless of merit, $TiO_2$, ZnO, AgNPs, or GNPs, all are not green pathway mediated products and may encompass toxicity and dangerous chemicals utilized in their manufacture. Green metallic NPs (MNPs) utilized in cosmetic formulations can easily enter the skin and interact with dermal protein molecules (elastin and collagen), as well as keratinocytes [4]. The discovery is significant in terms of both good and negative elements of nanotechnology. To begin with, for effective anti-wrinkle treatment in order to aim for cellular proteins such as elastin and collagen and boost their renaissance against aging, these MNPs would need to infiltrate the skin's intermediate coating (dermis). Second, because chemically generated MNPs carry hazardous substances, they can trigger gene mutations when they penetrate the skin. Another alarming point is the deposition of MNPs in other tissues or organs by passing through the skin barrier and may induce toxic conditions. Furthermore, releasing these MNPs into wastewater is harmful to the ecosystem. As a result, the idea of the green nano-based cosmeceuticals sector appears to be viewed solely in terms of good effects and is far from the actual implementation of green chemistry and nanotechnology's green potential. The evidence for these singularities is still being investigated. The Federal Drug Administration has also issued rules for the cosmetics industry's usage of nanotechnology [22].

## 12.3 Types of Green Nanomaterials

Biocompatibility, biodegradability, non-toxicity, cost-effectiveness, effortlessly programmable, and being ecological are some of the exclusive features of green nanomaterials created using green or sustainable techniques of synthesis [20].

i. ***Nanocellulose:*** In nature, the utmost plentiful green nanomaterial is nanocellulose being a copious nature-based polymer with lignocellulosic biomass as its noteworthy constituent. It is a natural nanomaterial consequential of plant cell walls. Among other qualities, it possesses high specific surface area, greater pliable module, low-density renewability, luxurious amalgamate, and highly stretchable forte [2]. Nanocellulose can be classified into three categories based on crystal structure, framework, particle dimensions, source, and scheme of abstraction: nanocrystalline cellulose (CNC), nano-fibrillated cellulose (CNF), and bacterial nanocellulose. Apart from the aforementioned characteristics, they also have a high water retention capacity, which allows hydrogels to be created and drug loading applications to be carried out. These hydrogels are also employed in wound healing because of their high permeability, ability to absorb exudates, and ability to stimulate healing when combined with various chemicals. They're employed in tissue engineering scaffolds because of their incredible tensile power and extremely hydrated three-dimensional assembly, which may precisely replicate the real tissues [3]. For example, CNCs, alginate composite with carboxymethyl chitosan aldehyde groups as a functional group of nanocellulose as an application application in wound dressing, CNCs, incorporated with poly vinyl alcohol (PVA) sulfate half-ester groups as an active group for wound dressing, CNFs, with carboxylic acid groups can also be utilized in wound dressing.

ii. ***Carbon Dots:*** These are being made from sustainable materials; earlier the synthesis approach was the utilization of diverse carbon-rich precursors origin of which was based on petroleum goods. However, restricted accessibility, greenhouse gas outcomes, underprivileged cost stability, adverse influence on climate, as well as the substitute of these resources were renewable feedstocks were particular shortcomings. This led to the utilization of this technique for synthesis involving an inclusive option for source choice, a copiousness of raw material, and eco-friendly options. The widespread range of applications of carbon dots (CDs) in diverse fields owes to the possession of its exclusive physico-chemical, functional, and tunable properties [20]. Due to their exceptional luminescent feature, chemical permanency, small toxicity, and less environmental threat, CDs have engrossed much attention. Some of its technological applications like optical detection, bioimaging, tribology, or photocatalysis, amid others, have been immensely sightseen. CDs are projected in the future to be pragmatic not just as green antioxidant-based extracts in bio-emollients, predominantly the ones based on vegetal oils, cosmetics, or in food packets but also in biodegradable lubes as green anti-friction and anti-wear nano-additives [16].

iii. ***Green Biocomposites:*** Materials developed from biological or renewable sources are known as green biocomposites and comprise reinforced and polymeric matrices. Polymers that are derived from natural sources can be handed-down as protecting mediators in the wrapping as they possess antioxidant and antibacterial action by adding nano-based clay and NPs in the polymer complex. When equated with their macro counterparts, greater surface-to-volume ratio and surface activity, as well as better biodegradability and biocompatibility, are all advantages of nanocomposites. Biocomposite NPs can be employed as a biodegradable polymer targetting arenas of vascular skin adhesive for treatment of injury, gene therapy as well as tissue engineering.

## 12.4 Synthesis

Green nanotechnology is referred to as synthesizing or manufacturing nanomaterials using the ethics of green chemistry to produce long-lasting, cost-effective, safe, and environmentally friendly nanomaterials. These synthetic materials are environmentally friendly and reduce the dangers of traditional NPs to human health. Green synthesis has received a lot of attention since it can make a wide range of nano-induced components, such as metallic or metallic oxide nano constituents, fusion materials, and biologically stimulated materials. The manufacture of nanomaterials can be categorized into two: top-down and bottom-up techniques. The constituent parts are cracked from their bulk counterpart up to a nanoscale level in a top-down approach, whereas the atoms are organized or coupled in detailed profiles to range up the level of nanoscale in a bottom-up approach. The physical, chemical, and green amalgamation are all methods for preparing nanomaterials. The physical synthesis of nanomaterials is mostly done from the top down. Chemical synthesis, on the other hand, takes a bottom-up method.

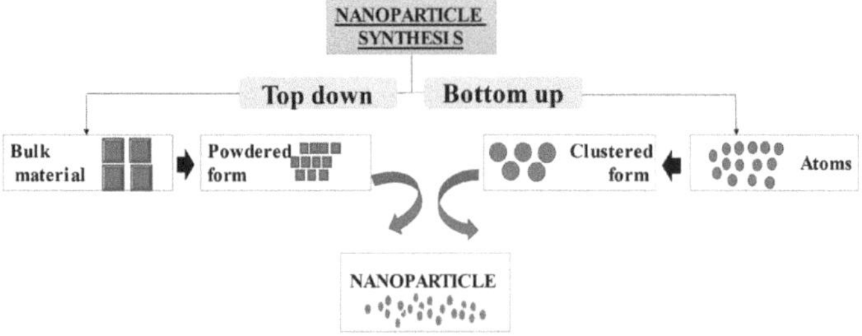


**Figure 12.2** Schematic representation of nanoparticle synthesis approach.

Incorporating the concept of green chemistry into the nanomaterial's synthesis process is another option [18]. Green chemistry is described as the use of a set of principles in designing, manufacturing, and implementing chemical products with the goal to reduce or eliminate the usage or spawning of perilous compounds. The underlying processes and mechanisms of "green" synthesis approaches have been studied in this regard. Plasma ionization, electrodeposition, sol-gel synthesis, chemical vapor deposition, ball milling, and the utilization of natural nano-based particles are some of the approaches for the creation of nanomaterial [18].

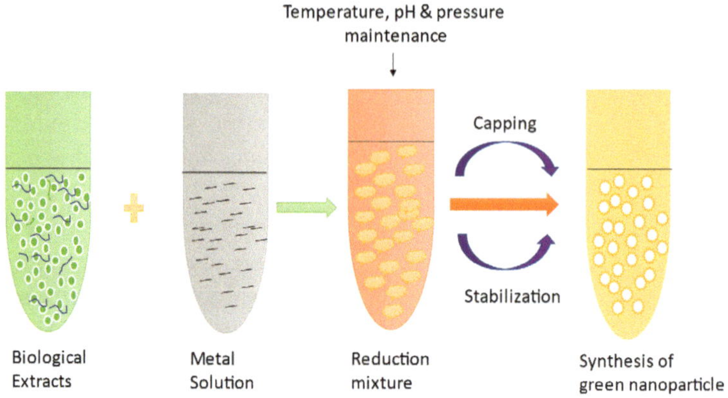


**Figure 12.3** Green nanoparticle synthesized using a biological method.

Green nanomaterials are created using natural reducing agents such as green solvents, biomolecules, and biopolymers. Supercritical fluids and amphiphilic solvents can also be used to make them [20]. Biological extracts are mixed with metal solution and a reduction reaction is carried out under optimum pH, temperature, and pressure followed by capping and stabilizing the solution. Thus, the green nanoparticle is synthesized [18]. The use of naturally produced compounds in the manufacture of multifunctional green nanomaterials is referred to as green biosynthesis of nanomaterials.

Green nanomaterials are made from a variety of proteins, plants, and microorganisms. The creation of NPs happens extracellularly in microorganisms, and the reduction of the substrate is accomplished by enzymes present in the microbe. Plants, on the other hand, include natural reducing and capping agents that aid in the creation of stable nanostructures. These nanomaterials offer a variety of features, like being biocompatible, biodegradable, eco-friendly as well as cost-effective, and can be effortlessly tweaked by changing their

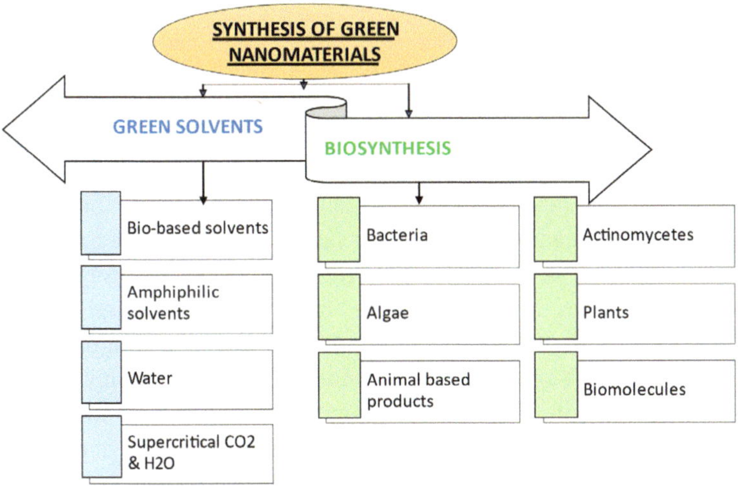


**Figure 12.4** Synthesis of green NPs.

process of synthesis [9]. To provide them supplementary benefits like huge surface area and added features nano scaling of these resources is performed. By altering the temperature and pressure of the solvent, supercritical, liquids having density, viscidness, and solvation properties transitional between the gas and liquid phases can be combined for synthesizing green nanomaterials. Due to their nonflammability and nontoxicity, supercritical carbon dioxide and supercritical water are acknowledged as green solvents. For the reason that they have less thickness, greater diffusion, and near about zero surface tension as their unique features, these are employed to prepare the nanocomposite. The surface qualities can be managed and altered to control and adjust the composite's geomorphology [18].

## 12.5 Action Mechanism

Skin covers the entire body and is considered as the prime tissue of the human body. It constitutes 3 sheets: the epidermis (the farthest sheet), the derma (hair sacs, perspiration secretors, and rough connective tissue are present in between), and the innermost hypodermis. The epidermis bifurcates into numerous coats and its outmost layer i.e., the stratum corneum, is accountable for the barricade purpose of the skin owing to its lipophilic nature and greater interconnection present amid the cells. Passive paths through which a particle can pass the stratum corneum could be intercellular (courses for drug around the skin), transcellular (through the skin cells), and appendageal (to bypass the stratum corneum) routes. Skin penetration of NPs occurs by the transportation of nanoparticulate through the skin. This happens due to the nature and physicochemical possessions of the nano elements and carrier used, the constituent actions as well as skin situations. However, dermal captivation of NPs occurs under certain circumstances. While cosmetic belongings aim to be applied on hail n hearty skin, they can also be applied on unusual skin. The blockade properties of skin may be lessened under such conditions. Studies advocate "nanoproducts applied onto the skin solitary breach via follicular openings hair and skin pores, and the trifling volume being traced underneath the stratum corneum". An investigation was conducted on the providence of these nano units based on cosmetic stuff. When tested onto mammalian tegument exploiting with laser scanning confocal microscopy to observe in case the fluorescently docketed constituent part (20–200 nm) were captivated hooked onto the skin revealed; nanoparticle communicating with unbroken or moderately injured skin cannot cross epidermis or beyond. With this, it is evidenced that the nano topical for cosmeceuticals distribution mechanisms is beneficial and harmless. Permeation (detection in the blood circulation) is undesired, but penetration and holding period in the epidermis or dermis is desirable when cosmetic products are concerned [5]. It is still noteworthy that nanoencapsulation improves the preservation of vegetable components in the epidermis and dermis. This finding is significant because it ensures the efficacy of naturally derived compounds loaded in nanocarriers, which is important given the current trend of adding natural ingredients into cosmetics. As revealed by confocal microscopy, nanoencapsulation of vegetable components modifies skin circulation and allows skin retention.

Cell viability, on the other hand, is a safety test, as stated in the previous quote. Because nano-systems may generate concerns about cell/tissue toxicity, determining their safety is critical. Low-cytotoxicity nanocarriers are regarded as safe. When compared to other efficacy assays, nanocarriers perform better. The antioxidant activity of vegetable components is increased when they are nano-encapsulated. Skin moisturization is also aided by nanocarriers, with the moisturizing impact varying depending on the carrier composition. Entrapped compounds' photostability is increased by nanocarriers, which impacts their antioxidant activity [10]. Because less of the chemical is damaged by light, more of it will be accessible to act as an antioxidant. Nano cosmetics are evaluated biologically with the similar assessment used for nano-based carters, such as skin penetration/permeation, hydration, and antioxidant activity. Other examinations include the sun protection factor and UV A protection factor, which are employed for semi-solid forms. Nanoparticle behavior must be studied further, especially whether NPs stay on the skin's surface are absorbed into the bloodstream, and reaches other organs.

## 12.6 Application in Cosmetics and Skin Treatment

People use cosmetics to polish their bodies, adorn their exteriors, and improve their appearance. As a result, cosmetics like creams, nail paint, hair colors, hair removers, conditioners, perfumes, body sprays, liniments, different powders, tooth cleansers, shaving emulsions, and antiperspirants are produced in large quantities. Almost all cosmetics companies and industries today use nanotechnology to create their goods.

Cosmetics are expected to be the fastest-growing segment of the personal care market. Lip, skin, nail, and hair products all have nanocosmeceuticals in them.

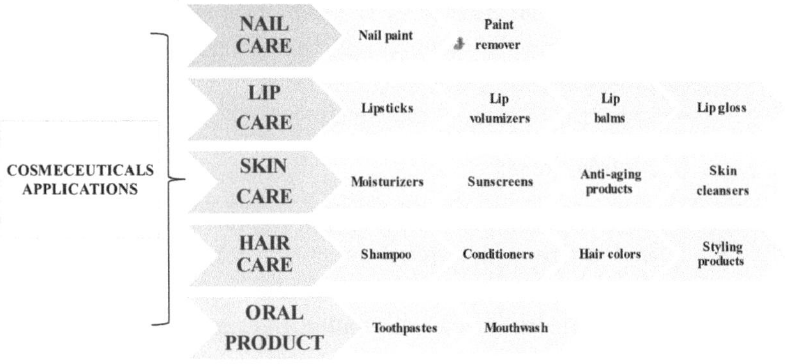


**Figure 12.5** Nanotechnology in cosmeceutical application.

The following are some of the most common types of nanocosmeceuticals [6]:

i. ***Nail care:*** Nanotechnology-based nail polishes provide advantages in terms of longevity, quick dry time, enhanced toughness, chip resistance, and ease of application due to their flexibility. Nano Labs Corp. received an interim patent for their unique nano nail rubdown and lacquer, which has the benefits of drying to a highly hard state, resisting stress, scratching, and chipping, and having a snap that allows for superior application without cracking. Incorporating NPs (silver and metal oxide-based) with antifungal properties in nail polishes for the treatment of fungal infections could be one of the emerging tactics with a lot of promise in the cosmeceutical industry [13].

ii. ***Lip care:*** It is another promising cosmeceutical category. According to a patent held by the Korean Research Institute of Bioscience and Biotechnology, it is possible to make pigments with a wide spectrum of hues by mixing NPs of gold or silver in constitutional proportions to preserve its color for a longer duration. Lip balm, stick, gloss, and volumizer are all available as nanocosmeceutical products. The various nanoparticles can be used in lipsticks and lip glosses to make them more flexible and hydrate the lips by reducing transepidermal water loss. Thus it helps in possessing the color intact for a longer stay by preventing pigment relocation from the lips [13].

iii. ***Skincare:*** These cosmetics are being used to advance the operational quality of the skin by promoting collagen enhancement also combating the detrimental effects of free radicals. They are also beneficial in dermis restitution, collagen revitalization, thus firming and lifting up the skin.

a. ***Moisturizers:***

For retaining moisture to improve the skin appearance, moisturizers are applied that forms a thin humectant film. Thus, by utilizing moisturizers that provide suppleness, skin dehydration can be avoided. Because of their extended benefits, nanoemulsions and solid lipid nanoparticles are commonly employed in moisturizing formulations.

They are also effective in the management of many skin disorders such as atopic dermatitis, psoriasis, and others [13].

b. ***Sunscreens:***

These are commonly used for skin protection from the detrimental consequences of sun's rays exposure. To protect the skin from being damaged by the sun, utmost permitted mineral-based constituents are ZnO and $TiO_2$. Sunscreens with ZnO or $TiO_2$ NPs being more transparent, less oily have a better visual appeal [13].

c. ***Antiaging Products:***

Skin aging is caused by daily exposure to chemicals, pollution, tension, irradiation from sources like infrared and UV, and abrasion. Collagen is an important component of skin regeneration and wrinkle reduction. The quantity of existing collagen in the skin decreases with aging.

Dryness, loss of elasticity and texture, the formation of spots, thinning, compromised barrier function, change in outward line isotropy, and, eventually leading to creases are all signs of skin aging. The majority of the cosmeceuticals developed claim to have anti-wrinkling, firming, moisturizing, and lifting properties, as well as whitening and skin harmonizing. The most common cosmeceuticals created using nanotechnology now in the market are skin rejuvenating products. Using nano techniques, L'Oreal has created Revitalift anti crinkle cream, which contains Pro-Retinol A nanosomes with the entitlements to decrease wrinkle look and retighten the skin instantly [13].

d. ***Skin Cleanser:***

Cornification and putrefaction products form a thin layer of film that encounters harmful microbes and acts as a natural barrier while also attracting dirt and pollutants from the surrounding. To preserve skin health, periodic cleansing is essential to eliminate debris, grime, and odor. Skin disinfectants and refiners are made from silver NPs. Nano Cyclic Inc. manufactures a scientifically harmonious balance of nanosilver and accepted chemical-based Nano

Cyclic cleanser pink soap, which professes to eliminate detrimental germs, combat acne, and minimize age spots as well as sun damage [13].

e. ***Haircare:***

Nanotechnology is being employed in hair care products, and ongoing research to learn how NPs can be incorporated to prevent hair loss and maintain hair health, luster, and silkiness. Conditioning agents, shampoos, hair growth accelerators, coloring, and styling treatments are among the hair products offered by nanocosmetics. Shampoo containing NPs seals dampness interior to cuticles by increasing inhabitant time of contact and thus forming a defensive coating between hair follicles and the scalp. Nanocosmeceutical ingredients in conditioning treatments impart gloss, silkiness, softness, and shine, as well as promote hair disentanglement. Cosmetics, unlike typical hair straightening products, do not disrupt the exterior structure of the hair fibers, known as cuticles, in order to penetrate into the hair strands [11].

f. ***Oral Products:***

Nanotechnology-based techniques have been employed in toothpaste, toothbrushes, mouthwashes, dental floss, and other oral care items to improve their efficacy over their traditional counterparts. Nanomaterials incorporated in toothpaste helps to improve tooth healing and facilitating demineralization thus rebuilding enamel, repairment of tooth sensitivity alongwith the fight assistance against infections caused by oral bacteria. Various NPs, such as gold, silver, zinc oxide, titania, and others, have been used in oral care. In a size-dependent way, strong antibacterial action has been exhibited by silver NPs against several aerobic and anaerobic oral pathogens such as *E. coli, A. actinomycetemcomitans, F. nuceatum, S. mitis, S. mutans, and S. sanguinis*. Because of their excellent antibacterial activity and biocompatibility with human cells, using silver based NPs in dental preparations can help manage microbial contaminations [22].

## 12.7 Challenging Aspects

Designing and expanding green nano-grounded ingredients, lifesaving nano-based pharmaceuticals, etc., in a resourceful, economical, and eco-friendly approach possess a massive trial. For producing nano underdone materials like MNPs, green synthesis can be a common podium for their miscellaneous applications. Although the ideologies of green chemistry are concerned with the approaches of green nano products and processes sturdily, however, the hands-on execution ratio is quite less [14]. Among all, one of the apparent causes could be technologies that have their nanophase expedition for improved performance underway; therefore, incorporating green principles can only be happening once these technologies are established completely [12].

### Case Studies:

- In a work by **Radmard et al** [19], Arbutin niosome (Arbusome); an ecological and green preparation revealed the skin permeation evaluations that exhibited arbutin deposits in a superior amount in the layers of skin, as well as cytotoxicity of augmented Arbusome in lipid nanotechnology for handling and control of hyperpigmentation complaints, was not revealed in MTT assay.
- **Morganti et al.** [15] experimented on producing an anti-aging safe and effective beauty mask. The material used in making masks was non-woven tissues using chitin-based nanofibril (CN) and nanolignin (LG) isolated from crustaceans and plant biomass respectively. For this initiative, electrospun fibers and NPs were characterized. The effectiveness and safety concerns of the produced masks were verified in vitro using corneocytes and fibroblasts culture being measured on 30 photo-aged women participants in vivo by expert dermatologists. The in vitro analysis exposed no toxicity on the feasibility of keratinocytes and fibroblasts when tested using the MTT technique. Produced beauty masks also articulated

reduction in cytokines expression, thus playing a key role in immunological and provocative responses regulation arising due to early aging and ecological assaults. The release of Metallo Proteinase I and the increased synthesis of collagen type I thus establishes the restituting and antiaging efficacy of produced pioneering beauty masks contributing to reduced skin aging. The initial primary outcomes in vivo, also confirm the shielding and invigorating action as revealed via the in-vitro learning escorted on 30 women volunteers signaling photoaging. These findings validate the potential of enhanced non-woven masks as an active topical mode of comforting itching or in the treatment of acne and rosacea skin ailments.

- **Conclusions:** In recent years, the increasing demand for bio-based cosmetics has drawn greater attention by both academia and the cosmetics industry to the extraordinary properties of green nanomaterials alongside their biodegradability, renewable character, and unique qualities. There was great interest in using NPs in the cosmetics, food, pharmaceuticals, agricultural and industries, focusing on more suited methods for the production of eco-friendly, non-toxic, and environmentally benign NPs using green biotechnology tools. The quality and amount of synthesized NPs for their potential use in different applications are affected by numerous factors. Toxic effects can be seen and must be thoroughly examined, such as dermal irritation, skin malignancy, and toxicity of genes. Experiments have led to the relations flanked by the nanomaterials and the skin. In cosmetic formulations, different types of NPs were used. However, additional studies are required in the direction to recognize the toxic consequences of nanomaterials, given the significant after-effect-free nature of beautifying products.

## References

1. Abbasi B.H., Fazal H., AhmadN., Ali M., Giglioli-Guivarch.N, Hano.C (2020) Nanomaterials for Cosmeceuticals: Nanomaterials-induced Advancement in Cosmetics, Challenges, and Opportunities; https://doi.org/10.1016/B978-0-12-822286-7.00005-X

2. Almeida T., Silvestre A.J.D., Vilela C., Freire C.S.R. (2014) Bacterial Nanocellulose toward Green Cosmetics: Recent Progresses and Challenges; https://doi.org/10.3390/ ijms22062836
3. Bacakova L., Pajorova J., Bacakova M., Skogberg A., Kallio P., Kolarova K., Svorcik V. (2019) Versatile Application of Nanocellulose: From Industry to Skin Tissue Engineering and Wound Healing; https://doi.org/10.3390/nano9020164
4. Brar S.K., Verma M., Zhang T., Das R.K., Tyagi R.D., Surampalli R.Y. (2015) Green Nanomaterials; https://doi.org/1061/9780784414088.ch23
5. Campbell C., Contreras-Rojas L., Delgado-Charro M., Guy R. (2012) Objective Assessment of Nanoparticle Disposition in Mammalian Skin after Topical Exposure; https://doi.org/10.1016/j.jconrel.2012.06.024
6. Dhawan S., Sharma P., Nanda S. (2020) Cosmetic nanoformulations and their intended use DOI: https://doi.org/10.1016/B978-0-12-822286-7.00017-6
7. Dureja H., Kaushik D., Gupta M., Kumar V., Lather V. (2005) Cosmeceuticals: An Emerging Concept; https://doi.org/37:1559.
8. Fytianos G., Rahdar A., Kyzas G.Z. (2020) Nanomaterials in Cosmetics: Recent Updates; https://doi.org/10.3390/nano10050979
9. Gottardo S., Mech A., Drbohlavova J., Małyska A., Bøwadt S., Sintes J.R., Rauscher J. (2021) Towards Safe and Sustainable Innovation in Nanotechnology: State-of-play for Smart Nanomaterials; https://doi.org/10.1016/j.impact.2021.100297
10. Júlia S.S. (2020) Nanocosmetics: Production, Characterization, and Performance Improvement; https://doi.org/10.5772/intechopen.93600
11. Kaul S., Gulati N., Verma D., Mukherjee S., Nagaich U. (2018) Role of Nanotechnology in Cosmeceuticals: A Review of Recent Advances; https://doi.org/10.1155/2018/3420204
12. DeLouise L.A. (2012) Applications of Nanotechnology in Dermatology; https://doi.org/10.1038/ jid.2011.425
13. Lohani A., Verma A., Joshi H., Yadav N., Karki N. (2014) Nanotechnology-Based Cosmeceuticals; http://dx.doi.org/10.1155/2014/843687
14. de la Guardia M. (2014) The Challenges of green nanotechnology, 4(1), 1–2 https://doi.org/10.5681/bi.2014.009
15. Morganti P., Palombo M., Carezzi F., Nunziata M.L., Morganti G., Cardillo M., Chianese A. (2016) Green Nanotechnology Serving the Bioeconomy: Natural Beauty Masks to Save the Environment; https://doi.org/10.3390/cosmetics3040041

16. Murru C., Badia L.R., Garcia M.E.D. (2020) Synthesis and Characterization of Green Carbon Dots for Scavenging Radical Oxygen Species in Aqueous and Oil Samples Antioxidants; https://doi.org/10.3390/antiox9111147

17. Nahar L., Sarker S.D. (2017) Importance of Nanotechnology in Drug Delivery; https://doi.org/1(5):000130.

18. Patra J.K., and Baek K.H., (2014) Green Nanobiotechnology: Factors Affecting Synthesis and Characterization Techniques; http://dx.doi.org/10.1155/2014/417305

19. Radmard A., Saeedi M., Morteza-Semnani K., Hashemi S.M.H., Nokhodchi A. (2021) An Eco-friendly and Green Formulation in Lipid Nanotechnology for Delivery of a Hydrophilic Agent to the Skin in the Treatment and Management of Hyperpigmentation Complaints: Arbutin niosome (Arbusome); https://doi.org/10.1016/j.colsurfb.2021.111616

20. Rawtani D., Rao P.K., Hussain C.M. (2020) Recent Advances in Analytical, Bioanalytical and Miscellaneous Applications of Green Nanomaterial; https://doi.org/https://doi.org/10.1016/j.trac.2020.116109

21. Salvioni L., Morelli L., Ochoa E., Labra M., Fiandra L., Palugan L., Prosperi D. Colombo M. (2021) The Emerging Role of Nanotechnology in Skincare; https://doi.org/10.1016/j.cis.2021.102437

22. Yadwade R., Gharpure S., Ankamwar B. (2021) Nanotechnology in Cosmetics Pros and Cons; https://doi.org/10.1088/2632-959X/abf46b

# Index